# Journal of Approximation Theory and Applied Mathematics

2013 Vol. 1 and Vol. 2

ISSN 2196-1581

Impressum:

Michael Rasguljajew

Kasinostraße 63
64293 Darmstadt
Germany, Hessen

E-Mail: m.rasguljajew@jatam.de

Website: www.jatam.com

Herstellung und Verlag:
BoD - Books on Demand, Norderstedt
ISBN 978-3-7357-9185-6

# Contents

## 2013 Vol. 1

*An Approximation on a Compact Interval Calculated with a Wavelet Collocation Method can Lead to Much Better Results than other Methods: 3 - 12*

*Parameter Identification with a Wavelet Collocation Method in a Partial Differential Equation: 13 - 19*

*An Approach for a Parameter Estimation with a Wavelet Collocation Method: 20 - 29*

*Notes on Nonparametric Regression with Wavelets: 30 - 39*

*Extrapolation and Approximation with a Wavelet Collocation Method for ODEs: 40 - 63*

## 2013 Vol. 2

*Solving ODEs and DAEs with a Wavelet Collocation Method with Examples from the Chemical Reaction Kinetics: 3 - 12*

*Solving Integral Equations with a Wavelet Collocation Approach: 13 - 16*

*Approximation of Non $L^2(R)$ Functions on a Compact Interval with a Wavelet Base: 17 - 24*

*Comparing Approximations of a Wavelet Collocation Method of Various Wavelets: 25 - 43*

# Journal of Approximation Theory and Applied Mathematics

2013 Vol. 1

## Contents

*An Approximation on a Compact Interval Calculated with a Wavelet Collocation Method can Lead to Much Better Results than other Methods: 3 - 12*

*Parameter Identification with a Wavelet Collocation Method in a Partial Differential Equation: 13 - 19*

*An Approach for a Parameter Estimation with a Wavelet Collocation Method: 20 - 29*

*Notes on Nonparametric Regression with Wavelets: 30 - 39*

*Extrapolation and Approximation with a Wavelet Collocation Method for ODEs: 40 - 63*

ISSN 2196-1581

# *An Approximation on a Compact Interval Calculated with a Wavelet Collocation Method can Lead to Much Better Results than other Methods*

M. Schuchmann and M. Rasguljajew from the Darmstadt University of Applied Sciences

### Abstract

As part of a research project we ran several simulations with a wavelet collocation method to find out how the optimal parameters can be determined. Comparing the approximations of functions on a compact interval $I$, we noticed that when $y$ is not in $L^2(R)$ a certain wavelet collocation method approximation was significantly better than projecting $1_I y$ orthogonal to $V_j$ (with the indicator function $1_I$). This method even gives very good approximations when using relatively few basis elements.

### Introduction

In the wavelet theory a scaling function $\phi$ is used, which belongs to a MSA (multi scale analysis). From the MSA we know, that we can construct an orthonormal basis of a closed subspace $V_j$, where $V_j$ belongs to a the sequence of subspaces with the following property:

$$\ldots \subset V_{-1} \subset V_0 \subset V_1 \subset \ldots \subset L^2(R),$$

$\{\phi_{j,k}(t)\}_{k \in Z}$ is an orthonormal basis of $V_j$ with $\phi_{j,k}(t) = 2^{j/2}\phi(2^j t - k)$.

We use the following approximation function

$$y_j(t) := \sum_{k=k_{min}}^{k_{max}} c_k \cdot \phi_{j,k}(t) \quad , \text{with } \phi \in C^1(R).$$

$k_{max}$ and $k_{min}$ depend on the approximation interval $[t_0, t_{end}]$ (see [7]).

Now we can approximate the solution of an initial value problem $y' = f(y,t)$ and $y(t_0) = y_0$ by minimization of the following function

(1) $$Q(c) = \sum_{i=1}^{m} \left\| y_j'(t_i) - f(y_j(t_i), t_i) \right\|_2^2 + \left\| y_j(t_0) - y_0 \right\|_2^2 .$$

For $m = |k_{max} - k_{min}|$ we get an equivalent problem:

$$y_j'(t_i) = f(y_j(t_i), t_i) \text{ for } i = 1, 2, \ldots, m \text{ and } y_j(t_0) = y_0.$$

Analogous we could treat boundary conditions instead of the initial condition. This method can be even used analogous for PDEs, ODEs of higher order or ODEs, which have the Form $F(y', y, t) = 0$.

## Error Estimation

For the orthogonal projection $y_j$ from $y$ in $V_j$ we know from the Gilbert-Strang Theory (see [9]) an upper bound of the approximation error in dependency of the order $p$: If the wavelet is of order $p$ then the approximation error has the order $O(2^{-jp})$ if $\|y^{(p)}\|_{L^2} < \infty$ and

$$\|y - y_j\|_{L^2} \leq C_\phi \cdot 2^{-jp} \cdot \|y^{(p)}\|_{L^2} .$$

If a wavelet is of order $p$ the scaling function $\phi$ even has a interpolation property, because then we can construct the functions $t^r$ with $r = 0, 1, ..., p-1$ over a linear combination of $\phi(t-k)$ (see [9]). That's also a property of the so called interpolating wavelets. For interpolating wavelets we find error estimations in [5] and [8].

In the examples we used the Shannon wavelet. For this wavelet we have additional information about the error in the Fourier space from the Shannon theorem (under the conditions of this theorem). For a good approximation with a small $j$ the behavior of $Y(\omega)$ with growing $|\omega|$ is important, because

$$y(t) - y_j(t) = \frac{1}{\sqrt{2\pi}} \int_{-\infty}^{\infty} Y(\omega) e^{i\omega t} d\omega - \frac{1}{\sqrt{2\pi}} \int_{-2^j\pi}^{2^j\pi} Y(\omega) e^{i\omega t} d\omega$$

$$= \frac{1}{\sqrt{2\pi}} \int_{-\infty}^{-2^j\pi} Y(\omega) e^{i\omega t} d\omega + \frac{1}{\sqrt{2\pi}} \int_{2^j\pi}^{\infty} Y(\omega) e^{i\omega t} d\omega .$$

With the Parseval theorem we get

$$\|y - y_j\|_{L^2} = \sqrt{\int_{-\infty}^{-2^j\pi} |Y(\omega)|^2 d\omega + \int_{2^j\pi}^{\infty} |Y(\omega)|^2 d\omega} .$$

But if we calculate $y_j$ over the minimization of $Q$ we generally don't get a orthogonal projection form $y$ in $V_j$ and generally $y$ is not quadratic integrable over $R$. There we can use the following theorem:

**Theorem 1:**
Assumptions: We have a initial value problem $y' = f(y, t)$ with $y(t_0) = y_0$ and

$$\|y_j(t_0) - y(t_0)\| \leq \delta ,$$

(4) $\quad \|y_j'(t) - f(y_j(t), t)\| \leq M$

and

(5) $\quad \|f(y(t), t) - f(y_j(t), t)\| \leq L \cdot \|y(t) - y_j(t)\|$ with $L > 0$.

Then we get for $t \geq t_0$:

$$\|y_j(t) - y(t)\| \leq \delta \cdot e^{L(t-t_0)} + M/L \cdot (e^{L(t-t_0)} - 1)$$

For a proof see [6].

In the examples and in many simulations we saw that $\delta$ was very small. If we assume $\delta = 0$ then we get the following error estimation (if we consider the compact interval $I = [t_0, t_{end}]$ and under the assumptions of theorem 1):

$$\|y_j - y\|_{L^2(I)} \leq M/L \cdot \sqrt{\int_{t_0}^{t_{end}} (e^{L(t-t_0)} - 1)^2 dt}$$

$$\leq M/L \cdot \sqrt{\frac{1}{2L} \cdot \left(-4e^{L(t_{end}-t_0)} + 3 + e^{2L(t_{end}-t_0)}\right) + t_{end} - t_0}$$

So if $M$ is very small then we can get very good approximations if $f$ is Lipschitz continuous.

## Comparing the Two Ways of Approximation

Now we want to approximate two functions in the following two examples, which are not quadratic integrable on $R$.

**Example 1:**
We begin with an approximation of the function $y(t) = e^{-t}$ on $I = [0, 1]$. $y$ is not in $L^2(R)$, but every on $I$ continuous function is in $L^2(I)$ or $1_I y$ (with indicator function $1_I$ of $I$) is in $L^2(R)$. So we set $k_{max} = -k_{min} = 20$ and we calculate an approximation function by an orthogonal projection from $1_I y$ on $V_1$. Therefore we calculate the coefficients of the approximation function over a scalar product:

$$c_k = \langle 1_{[0,1]} y, \phi_{j,k} \rangle = \int_0^1 y(t) \cdot \phi_{1,k}(t) dt$$

With the Shannon wavelet we get a worse approximation (dashed line is thee graph of $y$):

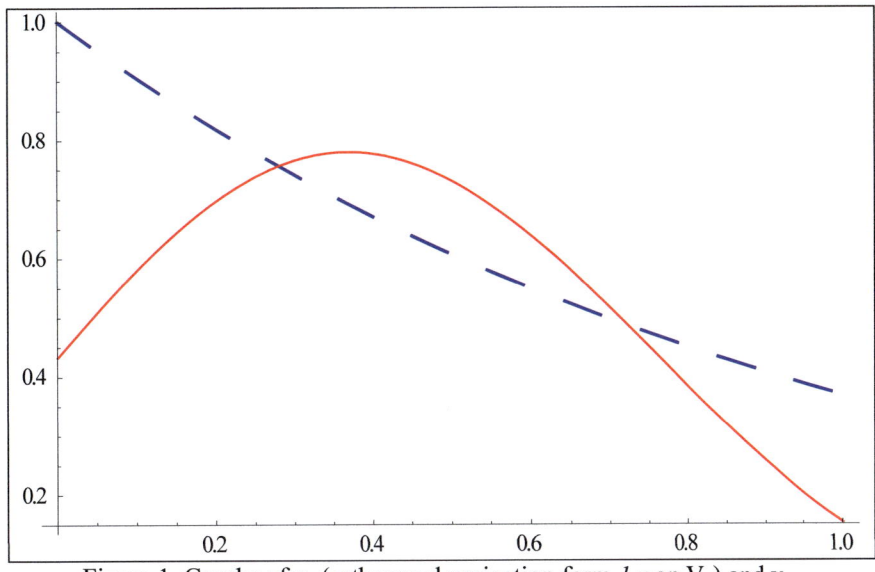
Figure 1. Graphs of $y_1$ (orthogonal projection form $I_1y$ on $V_1$) and y

With the Daubechies wavelet of order 8 we get no better approximation:

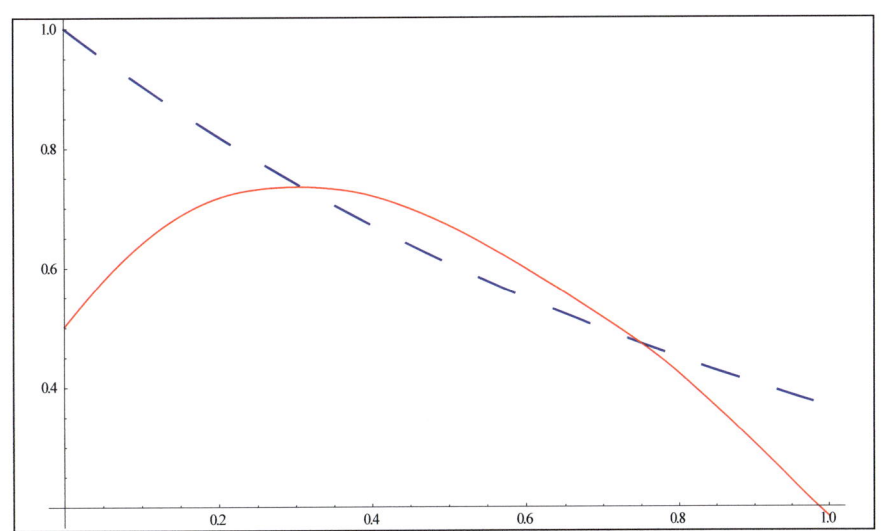
Figure 2. Graphs of $y_1$ (orthogonal projection form $I_1y$ on $V_1$) and y, Daubechies wavelet order 8

Even if we set $j = 3$ and $k_{max} =- k_{min} = 24$ we get not really a useful approximation:

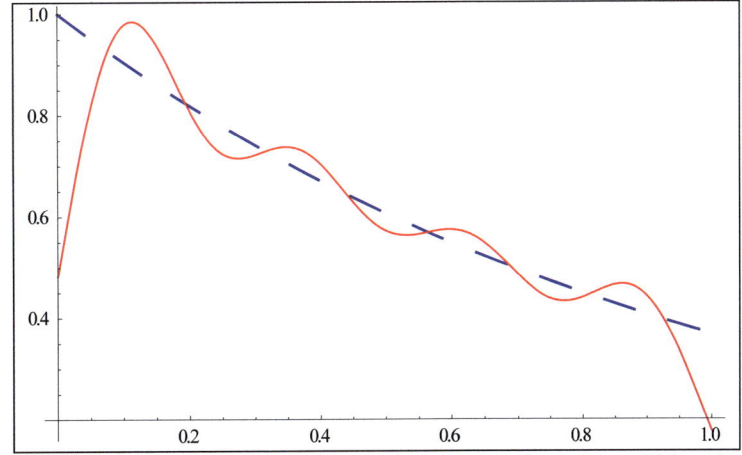
Figure 3. Graphs of $y_3$ (orthogonal projection form $I_1y$ on $V_3$) and y

If we take a look on the graphs on a bigger interval we see, that we calculated the best approximation of the function $1_{[0,1]} \cdot y$. That function is on $I$ identically to $y$ and outside $I$ equal to zero:

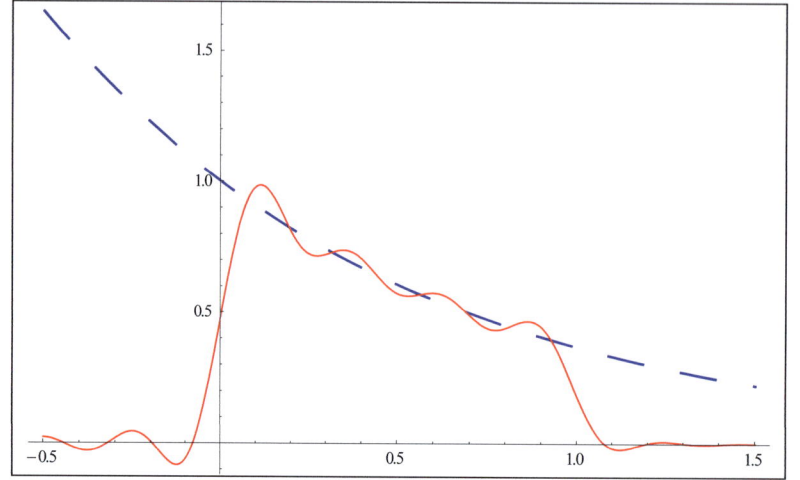

Figure 4. Graphs of $y_3$ (orthogonal projection form $1_I y$ on $V_3$) and y, bigger area

For a comparison the same approximation with the Daubechies wavelet of order 8:

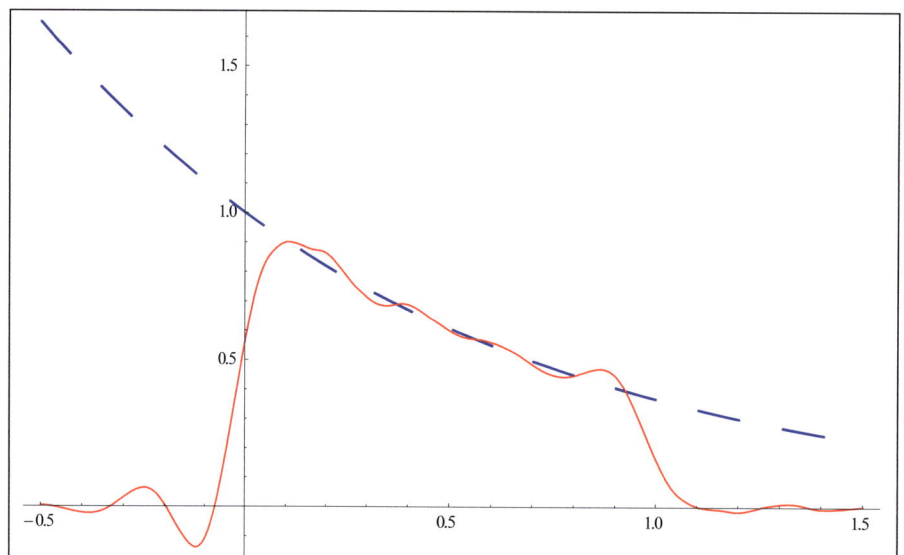

Figure 5. Graph of $y_3$ (orthogonal projection form $1_I y$ on V3) and y, Daubechies wavelet and bigger area

So the orhogonal projection considers the whole real axis, when we integrate only over $I$.

Now we calculate the coeffictions $c_k$ by the minimization of $Q$ (see (1)). We use the initial value problem $y' = -y$, $y(0) = 1$ and set an even smaller summation area with $k_{min} = -8$ and $k_{max} = 10$ and $j = 1$. We use the collocation points $t_i = i/20$ with $i = 0, 2, ..., 20$.

Graphically we see no difference between the approximation function $y_j$ and $y$ on $I$ ($Q_{min} \approx 3.00724 \cdot 10^{-30}$):

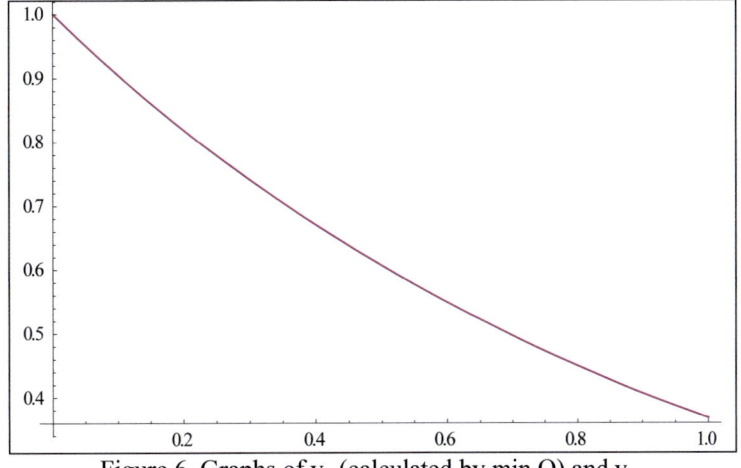
Figure 6. Graphs of $y_1$ (calculated by min Q) and y

Here is the graph of the difference function $y_j$ - $y$:

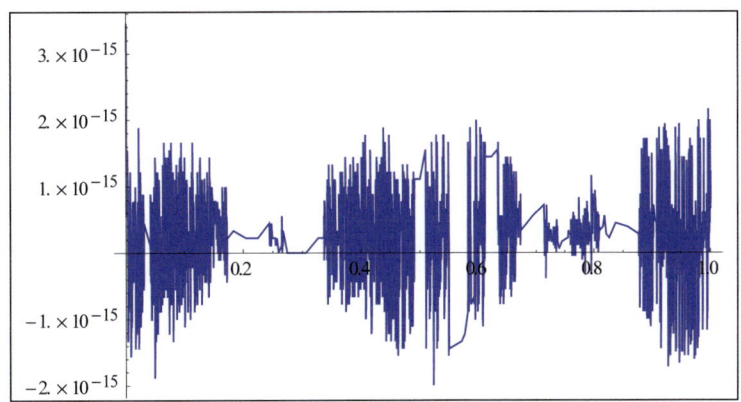
Figure 7. Graph of $y_1$ - y ($y_1$ calculated by min Q) and y

We could even use this approximation function for an extrapolation on a bigger interval than *I*. Here we see the graph of $y_1$ - y on the interval *[-0.5, 1.5]*:

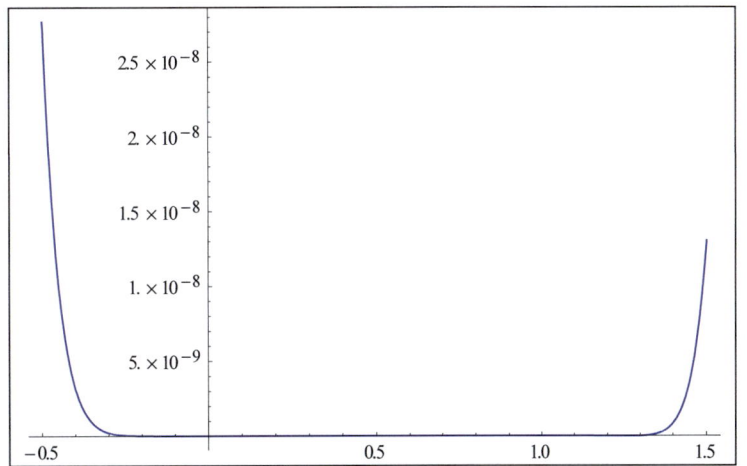
Figure 8. Graph of $y_1$ - y ($y_1$ calculated by min Q) and y on a bigger area

Here is the graph of *d* with $d(t) = (y_j(t) - f(y_j(t),t))^2$:

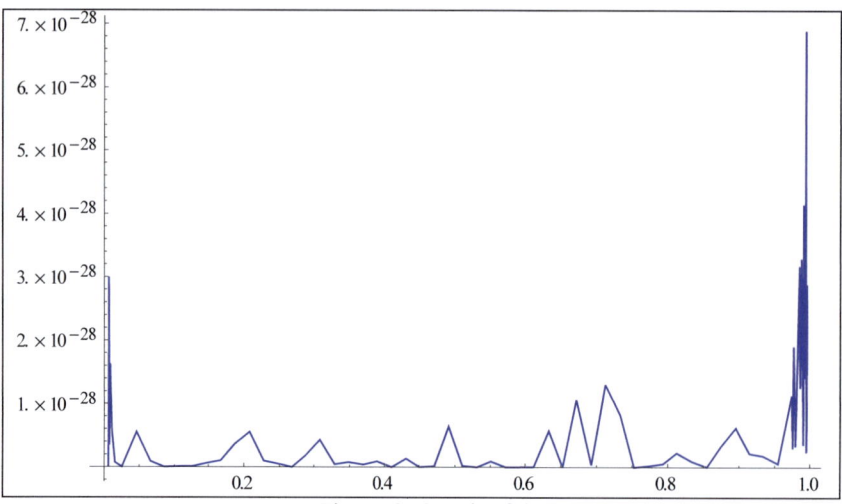

Figure 9. Graph of d

So *M* is small.

**Example 2:**
Now we consider the function $y(t) = sin(t)$, which is not quadratic integrable on $\mathbb{R}$ but we could construct $y$ with functions out of $V_j$ by using the Shannon wavelet.

If we use the Shannon wavelet, $y \in L^1(R) \cap L^2(R)$ is in $V_j$ if $supp\ Y \subseteq [-2^j \cdot \pi,\ 2^j \cdot \pi]$ (where $Y$ is the Fourier transform of $y$).

If we consider the function $h(t) = sin(at)$ then

$$H(\omega) = \frac{1}{\sqrt{2\pi}} \int_{-\infty}^{\infty} h(t) \cdot e^{-i\omega t} dt = i\sqrt{\pi/2} \cdot (\delta(\omega+a) - \delta(\omega-a)),$$

with the Dirac delta distribution $\delta$ (using for transformation and back-transformation the factor $1/\sqrt{2\pi}$). So the Fourier transform of $h(t) = sin(at)$ (from now we choose only $a > 0$) is not a function and $h$ is not quadratic integrable on $R$ but we can show that we get for the basis coefficients in Fourier space $<H, \Phi_{j,k}> = 2^{-j/2} h(k/2^j)$ for $a < 2^j \cdot \pi$.

So we set $c_k = 2^{-j/2} y(k/2^j)$ and we get for $k_{max} = -k_{min} = 15$ the following graph of $y_0 - y$:

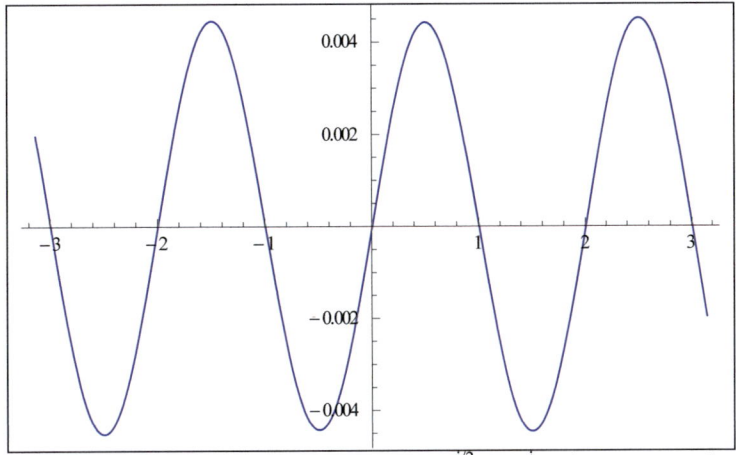
Figure 10. Graph of $y_0 - y$, $k_{max} = 15$, $c_k = 2^{-j/2} y(k/2^j)$, Shannon wavelet

For $k_{max} = -k_{min} = 50$:

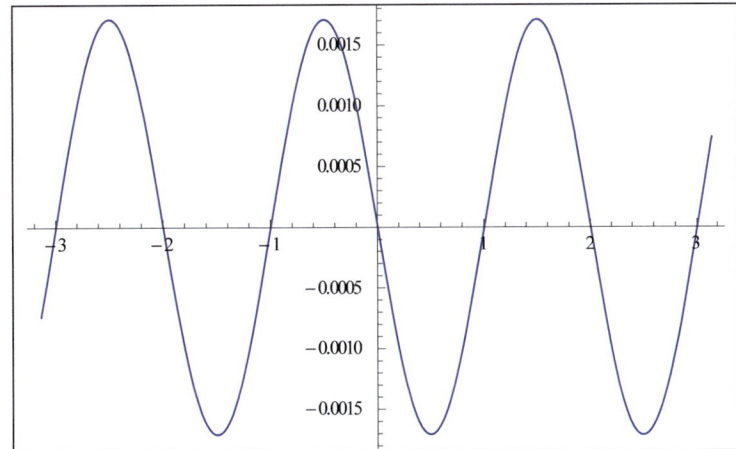
Figure 11. Graph of $y_0 - y$, $k_{max} = 50$, $c_k = 2^{-j/2} y(k/2^j)$, Shannon wavelet

For $k_{max} = -k_{min} = 100$:

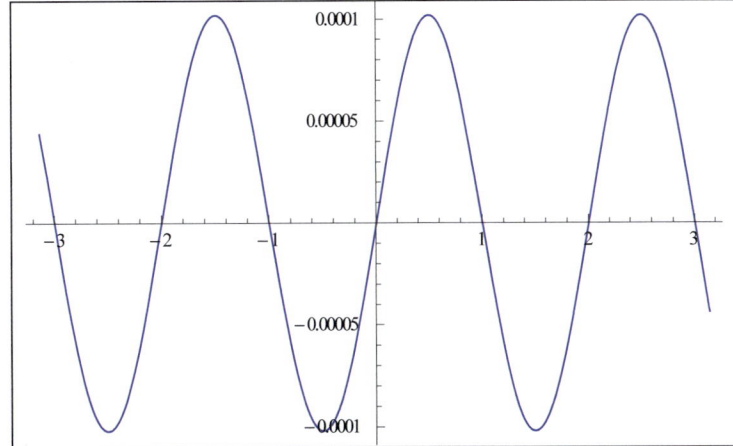
Figure 12. Graph of $y_0 - y$, $k_{max} = 100$, $c_k = 2^{-j/2} y(k/2^j)$, Shannon wavelet

If we would use the same method like in example 1 to get an approximation from $y$ on $I = [-\pi, \pi]$ and if we calculate $c_k = \langle 1_{[-5,5]} y, \phi_{k,j} \rangle$, so we get with $k_{min} = -k_{max} = 15$ and $j = 0$ a bigger error $y_0 - y$:

With the Shannon wavelet, we get the following difference $y_0 - y$:

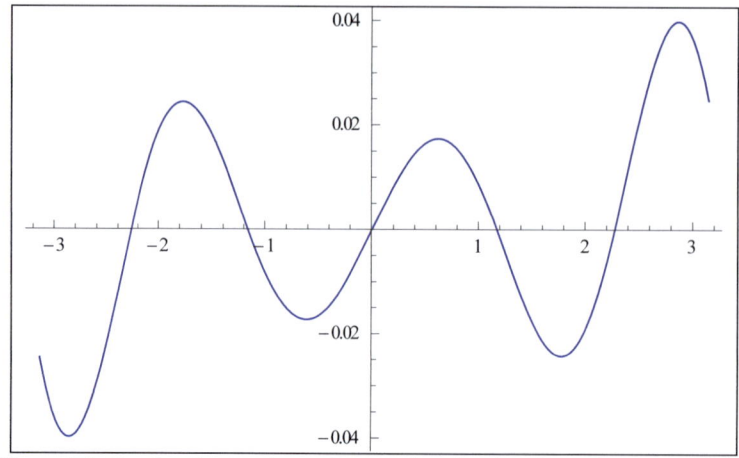

Figure 13. Graph of $y_0 - y$ ($y_0$ orthogonal projection from $1_{[-5,5]}y$ on $V_0$)

Here is the problem, that the Fourier transform of $y$ has a compact support, but not the Fourier Transform of $1_{[-5,5]}y$. Here ist the graph of magnitude spectrum of the Fourier transform of $1_{[-5,5]}y$:

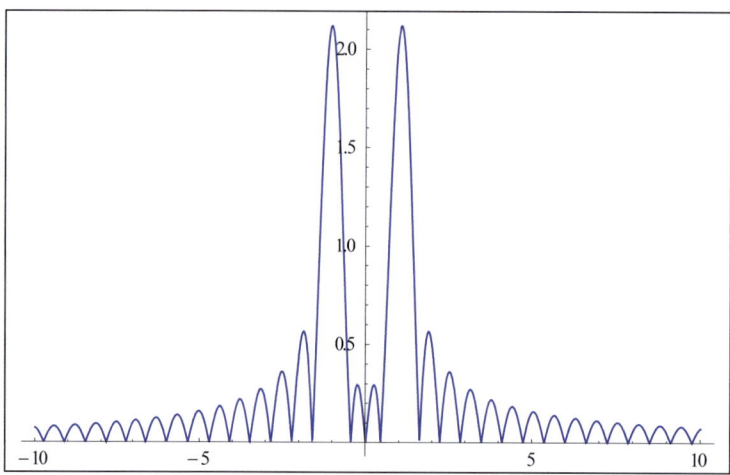

Figure 14. Magnitude spectrum of the Fourier transform from $1_{[-5,5]}y$

Now we consider the initial value problem with the solution $y(t) = sin(t)$:

$y'(t) = cos(t), y(0) = 0$.

We calculate the coefficients $c_k$ by minimization of $Q$ using the collocation points $t_j = -\pi + i \cdot \pi/15$, $i = 0, 1, ..., 30$. We set $j = 0$, $k_{max} = -k_{min} = 15$ an get the following difference $y_0 - y$ (with $Q_{min} \approx 4.7488 \cdot 10^{-28}$):

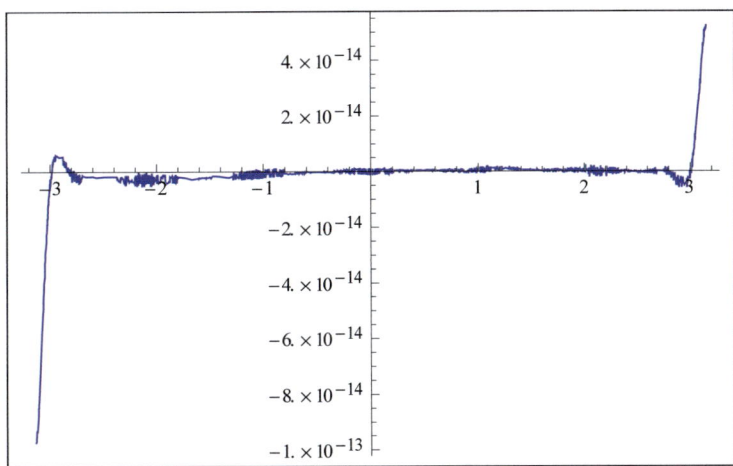

Figure 15. Graph of $y_0$ - y ($y_0$ calculated by min Q)

## Conclusions

If we use a wavelet basis for the approximation of a not quadratic integrablen function $y$ on a compact interval $I$, the calculation of an approximation function over the orthogonal projection form $I_j y$ on $V_j$ can lead to a worse approximation. But if we solve numerically an initial value problem with the solution $y$ by using a wavelet collocation method, we can get much better approximations. Even if $y$ would be band limited generally $I_j y$ is not band limited (because of the Heisenberg uncertainty principle in the Fourier transform)..

## References

[1] K. Abdella. *Numerical Solution of two-point boundary value problems using Sinc interpolation.* Proceedings of the American Conference on Applied Mathematics (American-Math '12): Applied Mathematics in Electrical and Computer Engineering, pp. 157-162, (2012).

[2] S. Bertoluzza. *Adaptive Wavelet Collocation Method for the Solution of Burgers Equation. Transport* Theory and Statistical Physics, 25:3-5, pp. 339-352, (2006).

[3] C. Blatter. *Wavelets – Eine Einführung.* 2nd edition, Vieweg, Wiesbaden, (2003).

[4] T. S. Carlson, J. Dockery, J. Lund. *A Sinc-Collocation Method for Initial Value Problems.* Mathematics of Computation, Vol. 66, No. 217, pp. 215-235, (1997).

[5] D. L. Donoho. *Interpolating Wavelet Transforms.* Tech. Rept. 408. Department of Statistics, Stanford University, Stanford, (1992).

[6] E. Hairer, G. Wanner. *Solving Ordinary Differential Equations I: Nonstiff Problems.* 2nd edition, Springer, Berlin, (1993).

[7] M. Schuchmann. *Approximation and Collocation with Wavelets. Approximations and Numerical Solving of ODEs, PDEs and IEs.* Osnabrück, DAV, *(2012).*

[8] Z. Shi, D.J. Kouri, G.W. Wie, D. K. Hoffman. *Generalized Symmetric Interpolating Wavelets.* Computer Physics Communications, Vol. 119, pp. 194-218, (1999).

[9] G. Strang. *Wavelets and Dilation Equations*: A Brief Introduction. SIAM Review, Vol. 31, No. 4, pp. 614-627, (1989).

[10] O. V. Vasilyev, C. Bowman. *Second-Generation Wavelet Collocation Method for the Solution of Partial Differential Equations.* Journal of Computational Physics, Vol. 165, pp. 660–693, https://wiki.ucar.edu/download/attachments/41484400/vasilyev1.pdf, (2000).

# *Parameter Identification with a Wavelet Collocation Method in a Partial Differential Equation*

M. Schuchmann and M. Rasguljajew from the Darmstadt University of Applied Sciences

## Abstract

In this article we describe a parameter identification method for an PDE. We use a wavelet collocation method and we show in a simulation that the error of the parameter estimation and of the approximation correlates with a sum of squares of the residuals. So we can assess the approximation function and the estimated parameters. This method can be applied analogous for PDEs of higher order. In the example we use the Shannon wavelet.

## Introduction of the Collocation Method

As an example we want to solve numerically a PDE (with boundary conditions) of first order

$$F(u(x,y), u_x(x,y), u_y(x,y), x, y) = 0,$$

$$u(x,0) = h(x)$$

on the area $D = [a_1, b_1] \times [a_2, b_2]$. Generally we use a scaling function $\phi \in C^r$ (if the order of the PDE is less or equal $r$). With that scaling function we can construct a two dimensional scaling function with

$$\phi(x,y) = \phi(x) \cdot \phi(y)$$

and we get the basis elements of $V_j$ with

$$\phi_{j,k_1,k_2}(x,y) = 2^j \phi(2^j x - k_1, 2^j y - k_2) \text{ and } k_1, k_2 \in Z.$$

With those basis elements we construct an approximation function (for easier notation we don't use the index $j$ in $g$):

$$g(x,y) = \sum_{k_1=n_u}^{n_o} \sum_{k_2=m_u}^{m_o} \phi_{j,k_1,k_2}(x,y) \cdot c_{k_1,k_2}$$

Now we have $n_k = (n_o - n_u + 1) \cdot (m_o - m_u + 1)$ unknown coefficients $c_{k_1,k_2}$.

To get an approximation of the solution $u$ we can solve the following equation :

(1a) $\quad F(g(x_i, y_l), g_x(x_i, y_l), g_y(x_i, y_l), x_i, y_l) = 0, \; i = 0, ..., n_1, l = 0, ..., n_2$

(1b) $\quad g(z_e, 0) = h(z_e), \; e = 0, ..., n_3$

With $(x_i, y_l) \in D$, $z_e \in [a_1, b_1]$, $n_k = (n_1+1) \cdot (n_2+1) + n_3 + 1$ and the collocation points $x_i \neq x_e$, $y_i \neq y_e$, $z_i \neq z_e$ for $i \neq e$.

Instead of the equations above we solve a minimum problem:

$$\min Q(c)$$

with

$$Q(c) = \sum_i \sum_l F(g(x_i, y_l), g_x(x_i, y_l), g_y(x_i, y_l), x_i, y_l)^2 + \sum_e (g(z_e, 0) - h(z_e))^2$$

This has the advantage that we can use more collocation points.

By using the same collocation points like in equations (1) the minimization problem is equivalent. If we use more collocation points in the minimum problem the equations (1) are only fulfilled approximately. But with a good approximation function $g$ the minimum of $Q$ is very small. If $F$ is linear then we have a quadratic minimization problem.

We use the collocation points:

$$x_i = a_1 + i \frac{b_1 - a_1}{n_1}, \quad i = 0, \ldots, n_1;$$

$$y_l = a_2 + l \frac{b_2 - a_2}{n_2}, \quad l = 0, \ldots, n_2;$$

$$z_e = a_1 + e \frac{b_1 - a_1}{n_3}, \quad e = 0, \ldots, n_3.$$

**Remark:**
A possible choice of the limits of the summation $n_o, n_u, m_o$ and $m_u$ is such that

(2) $$\phi_{j,k_1,k_2}(x, y) = 0$$

for $(x, y) \in D$ and $k_1 > n_o$, $k_1 < n_u$, $k_2 > m_o$, $k_2 < m_u$. If $\phi$ does not have a compact support or put generally we can replace (2) with

$$|\phi_{j,k_1,k_2}(x, y)| \leq \varepsilon$$

with a suitable $\varepsilon > 0$. So the limmits of summation depend on the approximation area D. In the example we will see that although the scaling function of the Shannon wavelet has not a compact support we need not many bases elements for a good approximation.

For the assessment of an approximation we compare $Q_{min} = \min Q(c)$ with

$$Q_a = \sum_i \sum_l F(g(\tilde{x}_i,\tilde{y}_l), g_x(\tilde{x}_i,\tilde{y}_l), g_y(\tilde{x}_i,\tilde{y}_l), \tilde{x}_i, \tilde{y}_l)^2 + \sum_e (g(\tilde{z}_e,0) - h(\tilde{z}_e))^2,$$

$$\tilde{x}_i = a_1 + i\frac{b_1 - a_1}{a \cdot n_1}, \quad i = 0,\ldots,n_1;$$

$$\tilde{y}_l = a_2 + l\frac{b_2 - a_2}{a \cdot n_2}, \quad l = 0,\ldots,n_2;$$

$$\tilde{z}_e = a_1 + e\frac{b_1 - a_1}{a \cdot n_3}, \quad e = 0,\ldots,n_3.$$

with an integer $a > 0$. Here $g$ is the approximation function with $c$ calculated by the minimization problem. In many simulations we got with $a = 2$ very good results like in example 1. The advantage of this method is that for $Q_a$ we don't have to calculate a second estimation because we get a whole approximation function $g$, not only points. If $Q_a$ is too big and $Q_{min}$ is very small we need more collocation points.

Analogous we can solve an PDE of higher order numerically with the described method, for example

$$F(u, u_x, u_y, u_{yy}, x, y) = 0,$$

$$u(x,0) = h_1(x) \text{ and } u_y(x,0) = h_2(x),$$

if we minimize

$$Q(c) = \sum_i \sum_l F(g(x_i,y_l), g_x(x_i,y_l), g_y(x_i,y_l), g_{yy}(x_i,y_l), x_i, y_l)^2$$
$$+ \sum_e (g_y(z_e,0) - x)^2 + \sum_e (g(z_e,0) - 0)^2$$

The same method we now apply to a parameter identification.

## Parameter Estimation and Assessment of the Approximation

We use as an example the following parameter identification problem with two parameters $p_1$ and $p_2$:

$$F(u, u_x, u_y, u_{yy}, x, y) = u_{yy} + p_1 u_y + p_2 u_x = 0,$$

$$u(x,0) = 0 \text{ and } u_y(x,0) = x.$$

Additionally we need measurements $\tilde{m}_{i,l}$ from $u$ at the points $(\hat{x}_i, \hat{y}_l)$.

We simulate the measurements and set $p_1 = 1$ and $p_2 = -1$, so that we get the exact solution of the PDE with boundary conditions:

$$u(x,y) = (x-2)(1 - Exp(-y)) + y \cdot (1 + Exp(-y)) \;.$$

We want to approximate the solution on the area $[0, 2]^2$.

Now $Q$ depends on two vectors $c$ and $p$ we minimize $Q(c, p)$:

$$\min_{c,p} Q(c, p)$$

with

$$Q(c,p) = \sum_i \sum_l F(g(x_i,y_l), g_x(x_i,y_l), g_y(x_i,y_l), g_{yy}(x_i,y_l), x_i, y_l)^2$$
$$+ \sum_e (g_y(z_e,0) - x)^2 + \sum_e (g(z_e,0) - 0)^2 + \sum_i \sum_l (\tilde{m}_{i,l} - g(\hat{x}_i, \hat{y}_l))^2$$

with $n_1 = n_2 = n_3 = 20$.

Using this method we even could identify parameters in the boundary conditions.

Now we calculate parameter estimations with $j = 0, 1, 2$ and $n_0 = m_0 = 2k_{max}$, $n_u = m_u = -k_{max}$ and $k_{max} = 3, 4, 5, 6$. As measurement points we choose $\hat{x}_i = i \cdot 1/8$, $i = 0,\ldots,16$, $\hat{y}_i = i \cdot 1/8$, $i = 0,\ldots,16$. We use the scaling function from the Shannon wavelet, so $\phi \in C^\infty$ (see [4]).

The estimated parameter we call $\hat{p}$. Here is the table of the results:

| j | $k_{max}$ | $Q_{min}$ | $Q_2$ | sse | $\|p - \hat{p}\|$ |
|---|---|---|---|---|---|
| 0 | 3 | 0.0000107237 | 0.000041234 | $7.70293 \times 10^{-9}$ | 0.0000561938 |
| 0 | 4 | $6.21836 \times 10^{-11}$ | $3.89477 \times 10^{-10}$ | $3.25975 \times 10^{-14}$ | $1.16275 \times 10^{-7}$ |
| 0 | 5 | $3.52194 \times 10^{-17}$ | $1.54002 \times 10^{-15}$ | $2.63439 \times 10^{-19}$ | $3.19512 \times 10^{-10}$ |
| 0 | 6 | $5.16334 \times 10^{-21}$ | $1.15253 \times 10^{-19}$ | $1.50842 \times 10^{-22}$ | $3.16654 \times 10^{-12}$ |
| 1 | 3 | 0.0730943 | 0.259421 | 0.00405652 | 0.00526086 |
| 1 | 4 | $5.40941 \times 10^{-7}$ | $3.05754 \times 10^{-6}$ | $4.44273 \times 10^{-10}$ | 0.0000121292 |
| 1 | 5 | $3.61066 \times 10^{-12}$ | $1.13727 \times 10^{-10}$ | $2.11763 \times 10^{-14}$ | $1.20021 \times 10^{-7}$ |
| 1 | 6 | $5.62028 \times 10^{-18}$ | $1.39483 \times 10^{-14}$ | $2.46281 \times 10^{-18}$ | $1.50996 \times 10^{-9}$ |
| 2 | 3 | 259.483 | 275.294 | 899.772 | 5.63328 |
| 2 | 4 | 144.504 | 235.327 | 325.062 | 3.60919 |
| 2 | 5 | 0.217185 | 2.11309 | 0.014662 | 0.0157611 |
| 2 | 6 | $4.21083 \times 10^{-8}$ | 0.0000177721 | $4.62554 \times 10^{-9}$ | 0.0000589473 |

$Q_{min}$ was calculated numerically by using the Mathematica function FindMinimum. $sse$ is the error sum of squares calculated with:

$$sse = \sum_{i=0}^{32} \sum_{j=0}^{32} (g(i/16, j/16) - g(i/16, j/16))^2$$

Here we see a correlation between $Q_{min}$ and $sse$, $Q_2$ and $sse$ and between $Q_2$ (or $Q_{min}$) and $sse$. In many simulations we saw that $Q_2$ is a better criterion to assess the estimation and the approximation because in $Q_2$ more points than the collocation points (with which we minimized $Q$) are considered. In many simulations we saw that in a bad approximation $Q_{min}$ can be small and $Q_2$ (or general $Q_a$) is relative big because the exact solution fulfils the PDE and the boundary value conditions at any point of the approximation area.

If we use a bigger $j$ then we need a bigger $k_{max}$ too because with rising $j$ we compress the bases functions $\phi_{j,k_1,k_2}$ .

In this example we have (as in other simulations we did) a linear dependency between $ln(Q_{min})$ and $ln(sse)$, between $ln(Q_2)$ and $ln(sse)$ and between $ln(Q_2)$ and $ln(||p - \hat{p}||)$.

This is what we see if we take a look at the regression tables:

**$ln(Q_{min})$ vs. $ln(sse)$:**

|   | Estimate | SE | TStat | PValue |
|---|---|---|---|---|
| 1 | -2.46106 | 1.22577 | -2.00776 | 0.0724447 |
| x | 1.05934 | 0.0502733 | 21.0715 | $1.28766 \times 10^{-9}$ |

$R^2$: 0.977974

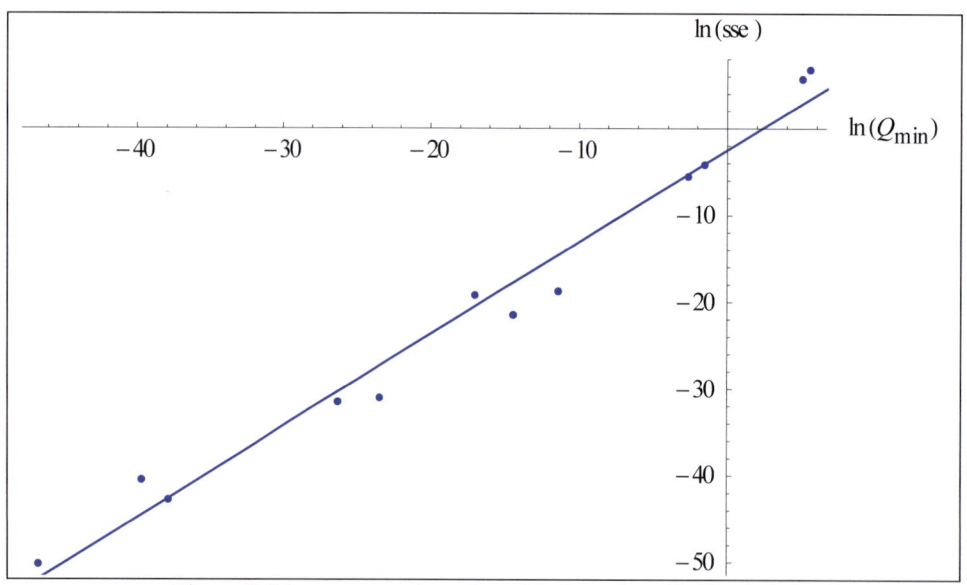

Figure 1. Linear regression plot with the points $(ln(Q_{min}), ln(sse))$

**$ln(Q_2)$ vs. $ln(sse)$:**

|   | Estimate | SE | TStat | PValue |
|---|---|---|---|---|
| 1 | -3.98319 | 1.11569 | -3.57015 | 0.00509447 |
| x | 1.15459 | 0.0520054 | 22.2014 | $7.71458 \times 10^{-10}$ |

$R^2$: 0.980115

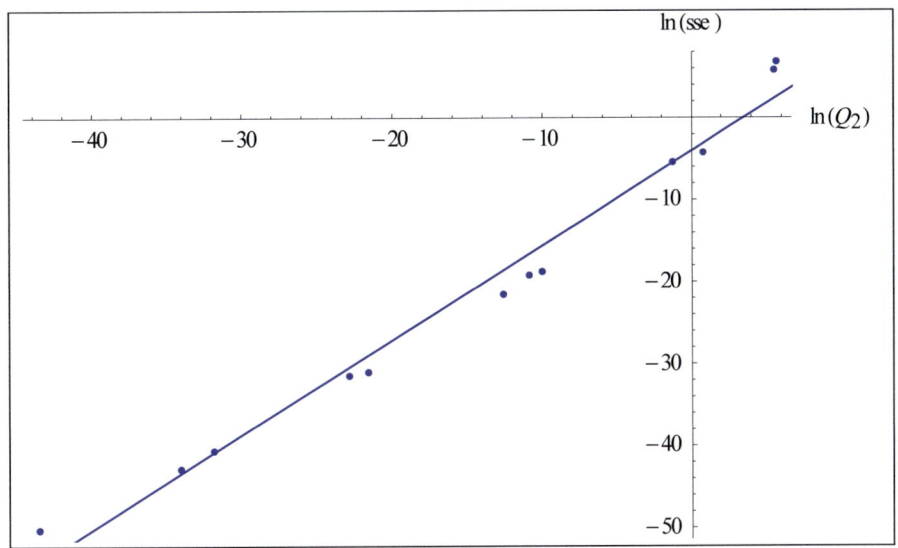

Figure 2. Linear regression plot with the points $(ln(Q_2), ln(sse))$

**$ln(Q_2)$ and $ln(\|p - \hat{p}\|)$:**

|   | Estimate | SE | TStat | PValue |
|---|---|---|---|---|
| 1 | -3.36543 | 0.453062 | -7.42819 | 0.0000224077 |
| x | 0.548946 | 0.0211184 | 25.9937 | $1.63369 \times 10^{-10}$ |

$R^2$: 0.985416

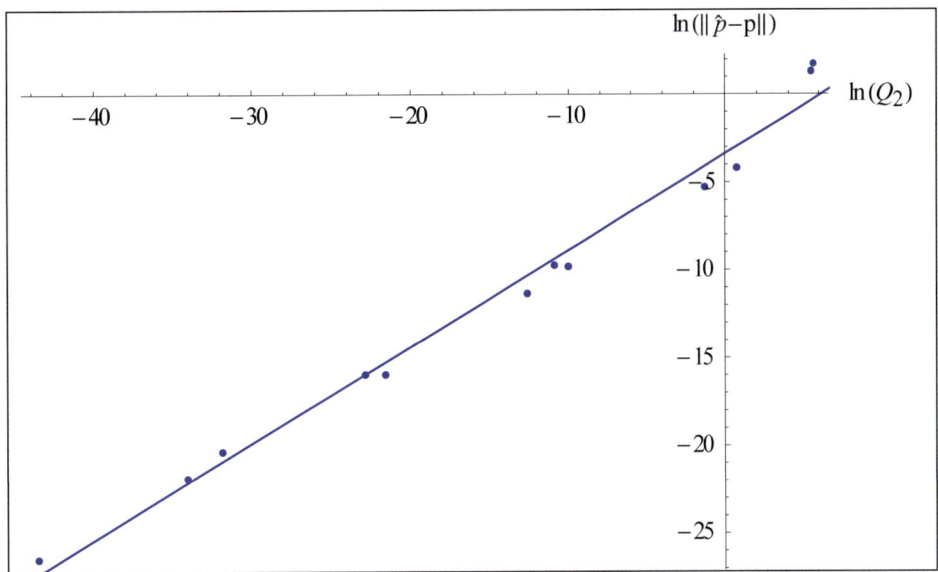

Figure 3. Linear regression plot with the points $(ln(\|\hat{p} - p\|), ln(sse))$

Here is the graph of $u - g$ for the best approximation (with the lowest $Q_2$ and $sse$):

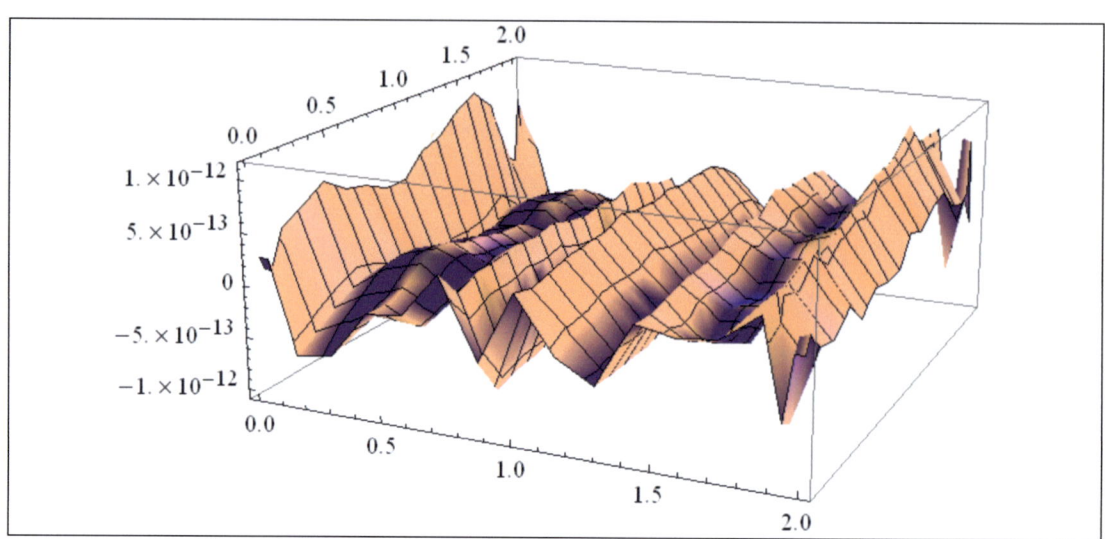

Figure 3. Graph of $u - g$ for the lowest $Q_2$ and $sse$

## Conclusions

The described wavelet collocation method can be used in different cases and different orders of the PDE (or ODE). The method can even be used if the boundary conditions include parameters and the PDE must not have a special form.

With $Q_{min}$ and $Q_2$ we can assess an approximation and the parameter estimation. If $Q_{min}$ is too big then the approximation function can not be used. On the other hand we can compare $Q_{min}$ with $Q_2$ and if $Q_2$ is too big ($Q_{min} << Q_2$) the approximation also can not be used. In both cases we need to calculate a new minimization. In the first case we do this by using another $j$ (or more bases functions if their number is too small). If we have enough basis functions then we can increase $j$. In the second case we need more collocation points.

## References

[1] T. S. Carlson, J. Dockery, J. Lund. *A Sinc-Collocation Method for Initial Value Problems.* Mathematics and Computation, Vol. 66, No. 217 (1997)

[2] S. Kameswaran, L. T. Biegler. *Simultaneous Dynamic Optimization Strategies: Recent Advances and Challenges.* Computers and Chemical Engineering, (2006).

[3] A. Nurmuhammada, M. Muhammada, M. Moria, M. Sugiharab. *Double Exponential Transformation in the Sinc-Collocation Method for a Boundary Value Problem with Fourth-Order Ordinary Differential Equation.* Journal of Computational and Applied Mathematics, (2005).

[4] L. Qian. *On the Regularized Whittaker-Koltel'nikov-Shannon Sampling Theorem.* Proceedings of the Amarican Mathematical Society, Vol. 131, No. 4, (2002)

[5] M. Schuchmann. *Approximation and Collocation with Wavelets. Approximations and Numerical Solving of ODEs, PDEs and IEs.* Osnabrück: DAV, (2012).

[6] M. Unser. *Vanishing Moments and the Approximation power of Wavelet Expansions.* Proceedings of the 1996 IEEE International Conference on Image Processing, (1996)

[7] R. Vuduc. *A Wavelet Collocation Method for Solving PDEs.* J. Comp. Phys., (2000).

# An Approach for a Parameter Estimation with a Wavelet Collocation Method

M. Schuchmann and M. Rasguljajew from the Darmstadt University of Applied Sciences

## Abstract

In this article we describe a parameter identification method for an ODE with an error approximation. We use a wavelet collocation method which can be applied to various problems like higher order ODEs, boundary value problems or PDEs. In the example we use the Shannon wavelet. In an example we apply this method on the pyridine problem form the reaction kinetics.

## Introduction

We use the following approximation function which is a function in the subspace $V_j$ of the $L^2(R)$:

$$y_j(t) := \sum_{k=k_{min}}^{k_{max}} c_k \phi_{j,k}(t)$$

$V_j$ is defined in the MSA (multi scale analysis). $\{\phi_{j,k}(t)\}_{k \in Z}$ is an orthonormal basis of $V_j$ with $\phi_{j,k}(t) = 2^{j/2} \phi(2^j t - k)$. $k_{max}$ and $k_{min}$ depend on the approximation interval $[t_0, t_{end}]$. For an ODE of order $r$ the scaling function $\phi$ should be in $C^r(R)$. If the ODE is a System, we use one approximation function for every component.

With the parameter identification we want to get the solution of

$$y'(t) = f(y(t), t, p)$$

for $p = p_{exact}$. The solution with the unknown parameter $p_{exact}$ we call $y_{p_{exact}}$. So the ODE depends on a parameter (vector) $p$ and we have additionally measurements $\tilde{y}_i$ of the function values $y_{p_{exact}}(\tilde{t}_i)$ at the measurement points $\tilde{t}_i$.

## A Classical Approach

A classical least square approach is to solve the following minimum problem

(1) $$\min \sum_{i=1,\ldots,\tilde{m}} \|\tilde{y}_i - y_p(\tilde{t}_i)\|^2.$$

Here and later $\|.\|$ is the Euklid norm. One possibility for a parameter identification with the wavelet collocation method is to start with a parameter $p_0$ and solve

$$y_j'(t_i) = f(y_j(t_i), t_i, p_0), \text{ with } i = 1, 2, \ldots, m$$
$$\text{and } y_j(t_0) = y_0, m = |k_{max} - k_{min}|.$$

Then we could apply a Gauss Newton step to solve numerically the minimum problem (1).

There we need additionally the numerically calculated derivates $\partial y_p(\tilde{t}_i)/\partial p$ at $p = p_0$ and we replace $y_p(\tilde{t}_i)$ through its linear approximation in the minimum problem (1). So we get $\hat{p}_1$ as a point at the minimum of the quadratic problem (because of the linearization), where we should additionally apply a descent test to get $p_1 = p_0 + \alpha(\hat{p}_1 - p_0)$.

## A Minimum Residual Approach with an Error Estimation

Another approach is the direct minimization of

$$Q_{\alpha,\beta}(p,c) = \alpha \cdot \sum_{i=0,\ldots,m} \|y_j'(t_i) - f(y_j(t_i), t_i, p)\|^2 + \beta \cdot \sum_{i=1,\ldots,\hat{m}} \|\tilde{y}_i - y_j(\tilde{t}_i)\|^2 + \|y_0 - y_j(t_0)\|^2.$$

$y_j(t_0) = y_0$ can be also used as a constraint. The numerical calculated value of the minimum from $Q_{\alpha,\beta}$ we call later $Q_{min}$ and we set $\alpha = \beta = 1$. The estimator of $c$ und $p$ we call $\hat{c}$ and $\hat{p}$.

To see, if the approximation function $y_j$ has also small residuals on other points than the collocation points, we compare $Q_{min}$ with

$$(2)\quad Q_{\alpha,\beta,a}(\hat{p},\hat{c}) = \alpha \cdot \sum_{i=0,\ldots,m_a} \|y_j'(\tau_i) - f(y_j(\tau_i), \tau_i, \hat{p})\|^2 + \beta \cdot \sum_{i=1,\ldots,\hat{m}} \|\tilde{y}_i - y_j(\tilde{t}_i)\|^2 + \|y_0 - y_j(t_0)\|^2$$

with $\tau_i = t_0 + i \cdot h/a$. $m_a = a \cdot m$ and an inter $a > 1$.

For the case $\alpha = \beta = 1$ we write short $Q_a$ (like $Q$ for $Q_{\alpha,\beta}$ in that case).

We already found further theoretical substantiations for $Q_a$ for the error estimation.

An estimation with less cost is, when we estimate in a first step $c$ and minimize $Q_{0,1}$:

(3a) $$Q_{0,1}(p,\hat{c}) = \min_c Q_{0,1}(p,c)$$

In a second step we estimate $p$:

(3b) $$Q_{1,0}(\hat{p},\hat{c}) = \min_p Q_{1,0}(p,\hat{c})$$

(3a) is only a quadratic problem and if $f$ is linear in $p$ (like in examples from the reaction kinetics) then (3b) is a quadratic problem, too. So we must only solve two times the normal equations. The cost for problem (3a) could be reduced even more if we chose special measurement points (see remarks 1 number 3) in the case that we have a small measurement error.

**Remark 1:**
$y_j$ depends on $c$. For easier notation we write short $y_j$. If we write $y_j$ outside a minimization problem depending on $c$, we mean that $y_j$ was calculated from (3a) and $c = \hat{c}$.

The estimation with less cost means:
In step 1 we minimize (for $c$)

$$\sum_{i=1,\ldots,\hat{m}} \left\| \tilde{y}_i - y_j(\tilde{t}_i) \right\|^2 + \left\| y_0 - y_j(t_0) \right\|^2 = \sum_{i=0,\ldots,\hat{m}} \left\| \tilde{y}_i - y_j(\tilde{t}_i) \right\|^2$$

with $\tilde{y}_0 = y_0$ and $\tilde{t}_0 = t_0$.

If

(4)
$$\min_c \sum_{i=0,\ldots,\hat{m}} \left\| y_{p_{exakt}}(\tilde{t}_i) - y_j(\tilde{t}_i) \right\|^2 = 0$$

(or we must consider the approximation error, see remarks 2) and (5) $\tilde{y}_i = y_{p_{exakt}}(\tilde{t}_i) + \varepsilon_i$, then

$$\sqrt{\min_c \sum_{i=0,\ldots,\hat{m}} \left\| \tilde{y}_i - y_j(\tilde{t}_i) \right\|^2} \leq \|\varepsilon\|$$

$\|\varepsilon\|$ can be estimated with the square root of $Q_{0,1}(p,\hat{c})$ if the measurement errors are independent and identically normal distributed (with mean 0).

In the second step we minimize (for $p$)

$$\sum_{i=0,\ldots,m} \left\| y_j'(t_i) - f(y_j(t_i), t_i, p) \right\|^2.$$

With the Gronwall Lemma we know, if

$y_{\hat{p}}'(t) = f(y_{\hat{p}}(t), t, \hat{p})$ with $y_{\hat{p}}(t_0) = y_0$ and

$$\|y_j(t_0) - y_{\hat{p}}(t_0)\| \leq \delta,$$

$$\|y_j'(t) - f(y_j(t), t, \hat{p})\| \leq M$$

and

$$\|f(y_{\hat{p}}(t), t, \hat{p}) - f(y_j(t), t, \hat{p})\| \leq L \cdot \|y_{\hat{p}}(t) - y_j(t)\|.$$

Then for $t \geq t_0$ we get:

$$\|y_j(t) - y_{\hat{p}}(t)\| \leq \delta \cdot e^{L(t-t_0)} + M/L \cdot (e^{L(t-t_0)} - 1)$$

In many simulations $y_j(t_0)$ is near $y_0$. For easier notation we assume $y_j(t_0) = y_0$, so we get

(6)
$$\sum_{i=0,\ldots,\hat{m}} \|y_j(\tilde{t}_i) - y_{\hat{p}}(\tilde{t}_i)\|^2 \leq (M/L)^2 \cdot \underbrace{\sum_{i=0,\ldots,\hat{m}} (e^{L(\tilde{t}_i - t_0)} - 1)^2}_{:=C_L}.$$

Generally we can use $\hat{M}_a^2 = \max \|y_j'(\tau_i) - f(y_j(\tau_i), \tau_i, \hat{p})\|^2$, $\tau_i = t_0 + i \cdot h/a$ and $i = 0, 1, \ldots, a \cdot m$ (with an integer $a > 1$) for an approximation of $M^2$ (compare with (2)). Here we know the following relation:

$$\hat{M}_a^2 \leq Q_a$$

With (4), (5) and (6) we get:

(7) $$\sqrt{\sum_{i=0,\ldots,\tilde{m}} \|y_{P_{exact}}(\tilde{t}_i) - y_{\hat{p}}(\tilde{t}_i)\|^2} \leq \|\varepsilon\| + M/L \cdot C_L$$

**Remarks 2:**
1) Form (4) and (5) follows

$$\sqrt{\sum_{i=0,\ldots,\tilde{m}} \|y_{P_{exakt}}(\tilde{t}_i) - y_j(\tilde{t}_i)\|^2} \leq \|\varepsilon\|$$

what we used for (7).

2) The assumption (4) is fulfilled for example,

a) if $\tilde{m} = |k_{max} - k_{min}|$. With that choice for the parameters $\tilde{m}$ we interpolate the measurements. This is only advisable if the measurement error $\varepsilon$ is small.

b) if $\phi$ has compact support, $y_{P_{exakt}}$ is in $V_j$ and if $k_{min}$ is small and $k_{max}$ is big enough so that $\phi_{j,k}(t) = 0$ for $t \notin [\tilde{t}_0, \tilde{t}_{\tilde{m}}]$ and $k < k_{min}$ or $k > k_{max}$.

c) if we use the Shannon wavelet and $\tilde{t}_i = 2^{-j} \cdot i$. Here:

$$\phi_{j,k}(2^{-j} \cdot i) = 2^{j/2} \cdot \phi(i-k) = \begin{cases} 2^{j/2} & \text{if } i = k \\ 0 & \text{else} \end{cases}$$

So $c_i = 2^{-j/2} \cdot y(2^j \cdot i)$ and $y(\tilde{t}_i) = y_j(\tilde{t}_i)$ with $y = y_{P_{exakt}}$. This is an interpolation property of the Shannon wavelet. If the measurement error $\varepsilon$ is small, we can use $\tilde{y}_i$ for an approximation of $c_i$. If $y$ is in $V_j$ then $c_i = 2^{-j/2} \cdot y(2^j \cdot i)$ are the exact bases parameters (follows from Shannon's theorem).

3) The assumption (4) can be not fulfilled if we use arbitrary $\tilde{t}_i$ and $y$ is not in $V_j$ or if the scaling function $\phi$ has no compact support and we choose a too less basis elements $\phi_{j,k}$ for the calculation for $y_j$. In the first case we have to consider the aliasing error and in the second case we have to consider the error depending on $k_{min}$ and $k_{max}$ caused through the missing bases coefficients (for a too big $k_{min}$ or a too small $k_{max}$).

In both cases we have to consider the approximation error $\tilde{\varepsilon}$ (i.e. the term $\|\tilde{\varepsilon}\|$) in the equation (7), if

$$y_{P_{exakt}}(\tilde{t}_i) - y_j(\tilde{t}_i) = \tilde{\varepsilon} \quad \text{with } c = \hat{c} \text{ from } \min_c \sum_{i=0,\ldots,\tilde{m}} \|y_{P_{exakt}}(\tilde{t}_i) - y_j(\tilde{t}_i)\|^2 = \|\tilde{\varepsilon}\|^2.$$

# Parameter Identification with the Pyridine Problem

The reaction scheme for the denitrogenation of pyridine is as follows:

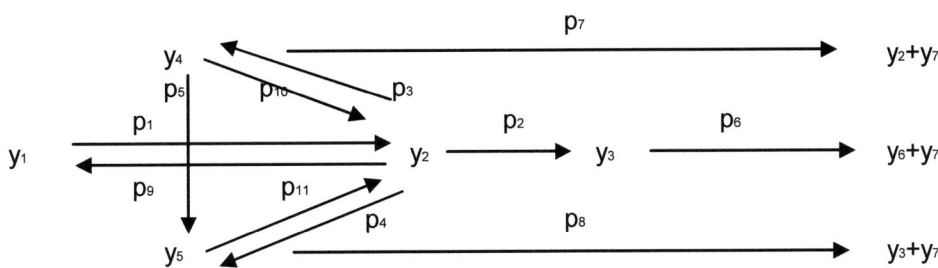

Figure 1. Reaction scheme pyridine

This yields to the following system of differential equations for the pyridine problem:

$$y_1' = -p_1 y_1 + p_9 y_2$$
$$y_2' = p_1 y_1 - p_2 y_2 - p_3 y_2 y_3 + p_7 y_4 - p_9 y_2 + p_{10} y_4 y_6$$
$$y_3' = p_2 y_2 - p_3 y_2 y_3 - 2 p_4 y_3 y_3 - p_6 y_3 + p_8 y_5 + p_{10} y_4 y_6 + 2 p_{11} y_5 y_6$$
$$y_4' = p_3 y_2 y_3 - p_5 y_4 - p_7 y_4 - p_{10} y_4 y_6$$
$$y_5' = p_4 y_3 y_3 + p_5 y_4 - p_8 y_5 - p_{11} y_5 y_6$$
$$y_6' = p_3 y_2 y_3 + p_4 y_3 y_3 + p_6 y_3 - p_{10} y_4 y_6 - p_{11} y_5 y_6$$
$$y_7' = p_6 y_3 + p_7 y_4 + p_8 y_5$$

The following results are related to the initial vector

$$y(0) = (1,0,0,0,0,0,0)^T .$$

In this reaction, pyridine is transferred with the help of three catalysts in ammonia and pentane (see [3]), which takes about 5.5 seconds. This reaction takes place isothermally at 350° K under a pressure atm of 100.

It was used the parameter vector

$$p = (1.81, 0.894, 29.4, 9.21, 0.058, 2.43, 0.0644, 5.55, 0.0201, 0.577, 2.15)^T .$$

Now we come to the parameter estimation using the approximation $y_j$ ($k_{max}$ = 20 and $k_{min}$ = -5).

We set $j = 1$, choose $I = [0, 5]$ (d.h. $t_{end} = 5$) and as collocation points we use

$$t_i = 1/16 \cdot i, \text{ with } i = 1,...,80$$

($m = 80$). As measurement points we used $\tilde{t}_i = 0.1 \cdot i$, with $i = 1,...,50$ ($\tilde{m} = 50$).

So:

$$Q_{\alpha,\beta}(p,c) = \alpha \cdot \sum_{i=1,\ldots,80} \|y_j{'}(t_i) - f(y_j(t_i), t_i, p)\|^2 + \beta \cdot \sum_{i=1,\ldots,50} \|\tilde{y}_i - y_j(\tilde{t}_i)\|^2 + \|y(t_0) - y_0\|^2$$

At first we do an estimation in two steps as described as follows

First Step:
$$Q_{0,1}(p, \hat{c}) = \min_c Q_{0,1}(p, c)$$

Second Step:
$$Q_{1,0}(\hat{p}, \hat{c}) = \min_p Q_{1,0}(p, \hat{c})$$

We had to solve two times a quadratic problem.

Here are the results of the first step:

$$Q_{0,1}(p, \hat{c}) = \min_c Q_{0,1}(p, c) \approx 2.99612 \cdot 10^{-9}$$

Here are the graphs of $y_1^{(i)} - y^{(i)}$ beginning with $i = 1$ ($y$ is here the numerically calculated function using the Mathematica function NDSolve):

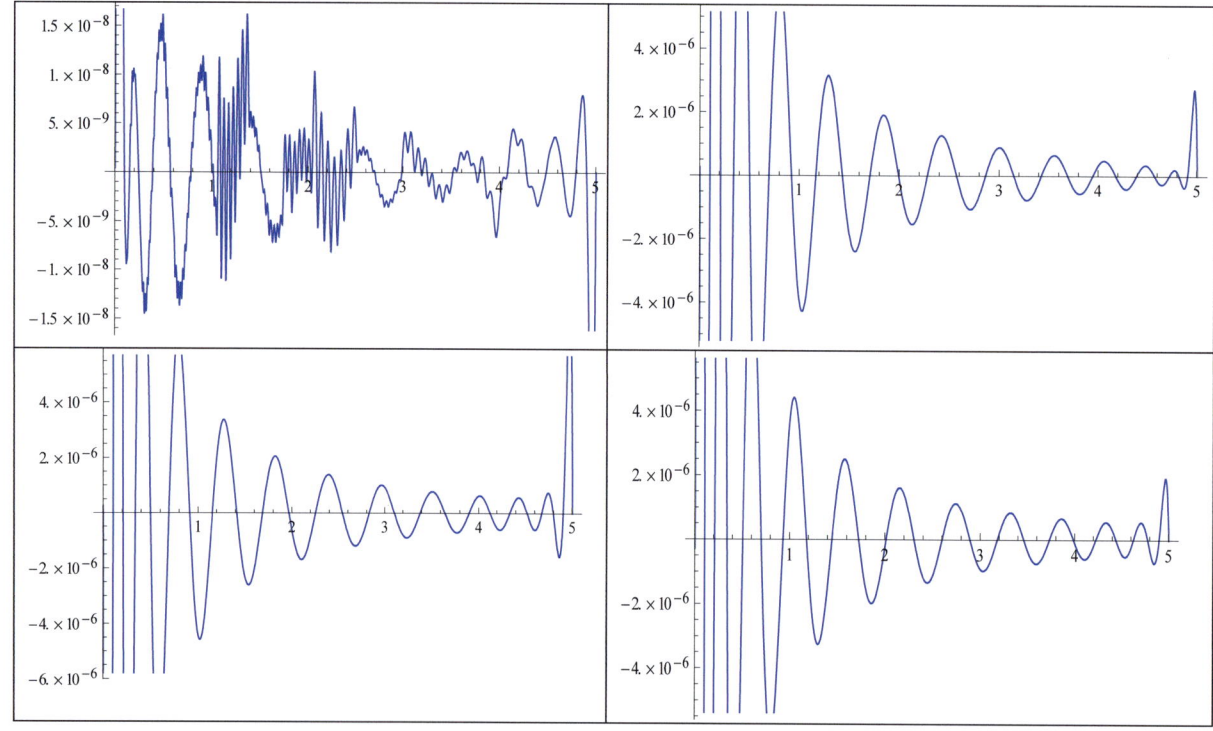

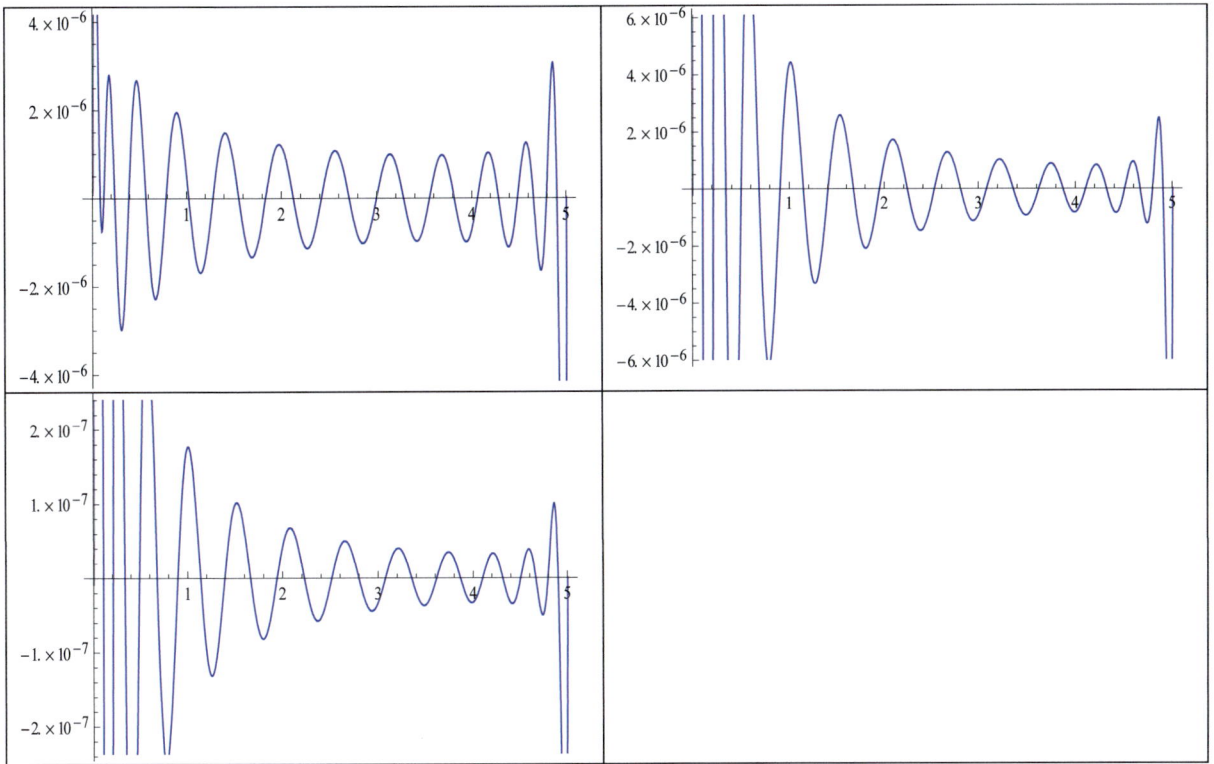

Figure 2. Graphs of $y_1^{(i)} - y^{(i)}$, top left for $i = 1$, to the right for $i = 2$, etc.

Here are the in the second step estimated parameters:

| $p_i$ | $\hat{p}_i$ | $|(\hat{p}_i - p_i)/p_i|$ |
|---|---|---|
| 1.81 | 1.81125 | 0.00069058 |
| 0.894 | 0.894442 | 0.000494432 |
| 29.4 | 29.4016 | 0.0000557119 |
| 9.21 | 9.38508 | 0.0190099 |
| 0.058 | 0.0576356 | 0.00628269 |
| 2.43 | 2.42631 | 0.00151957 |
| 0.0644 | 0.0651214 | 0.0112015 |
| 5.55 | 5.53804 | 0.00215583 |
| 0.0201 | 0.0209635 | 0.0429597 |
| 0.577 | 0.576463 | 0.000929825 |
| 2.15 | 2.22964 | 0.0370417 |

We got: $Q_{1,0}(\hat{p},\hat{c}) = \min_{p} Q_{1,0}(p,\hat{c}) \approx 1{,}18486 \cdot 10^{-4}$

Now again both parameters were estimated together (with $\alpha = \beta = 1$). The following was found:

$$\min_{c,p} Q_{1,1}(p,c) \approx 6.87561 \cdot 10^{-6}$$

For a comparison, $Q_2 \approx 4.92186 \cdot 10^{-4}$.

Here are the graphs of $y_1^{(i)} - y^{(i)}$ beginning with $i = 1$:

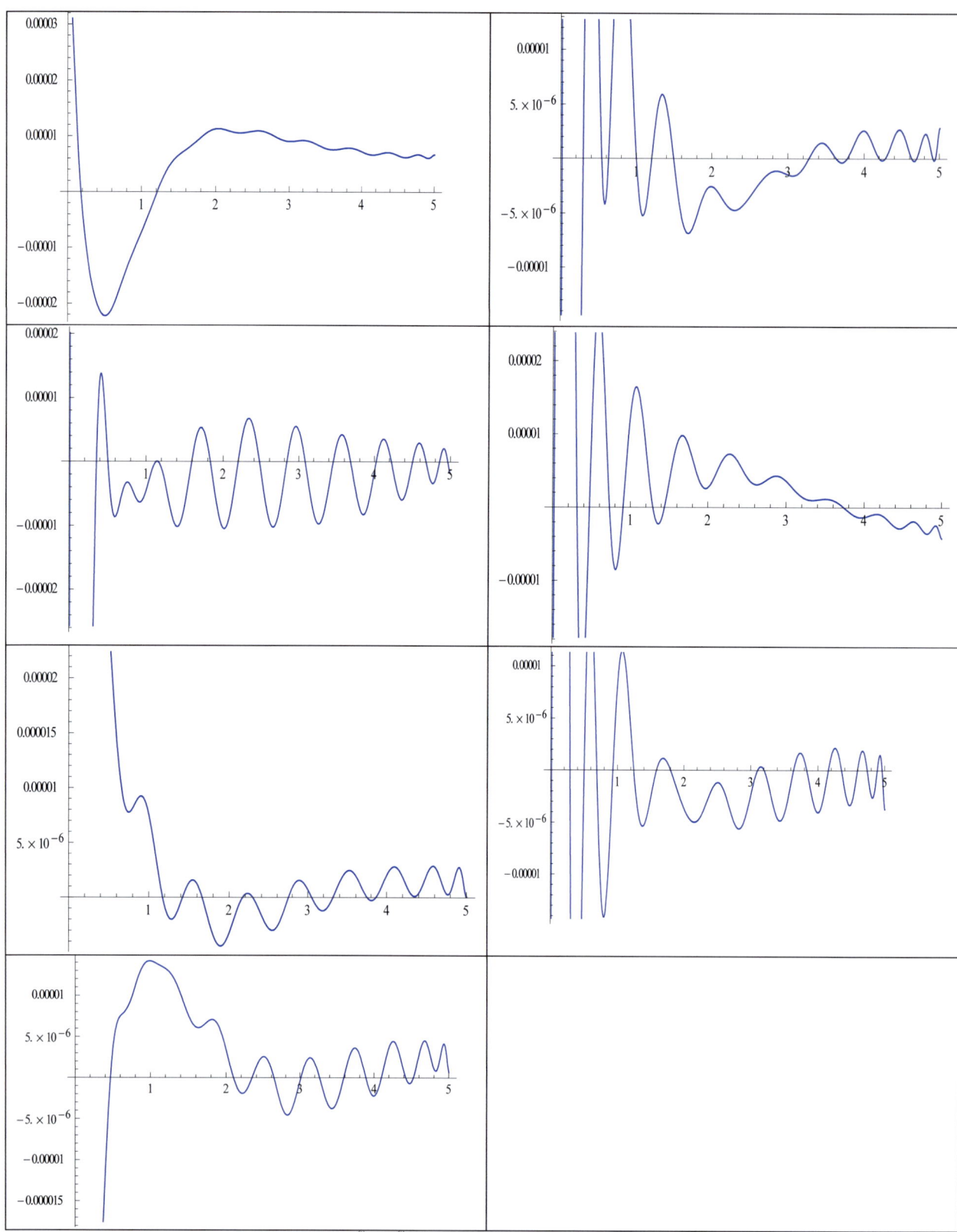

Figure 3: Graphs of $y_1^{(i)} - y^{(i)}$, top left for $i = 1$, to the right for $i = 2$, etc.

Here is the parameter estimation for $p$:

| $p_i$ | $\hat{p}_i$ | $|(\hat{p}_i - p_i)/p_i|$ |
|---|---|---|
| 1.81 | 1.81033 | 0.000180172 |
| 0.894 | 0.893921 | 0.0000879803 |
| 29.4 | 29.4036 | 0.000124069 |
| 9.21 | 9.36569 | 0.0169047 |
| 0.058 | 0.0580298 | 0.000514097 |
| 2.43 | 2.42799 | 0.000827474 |
| 0.0644 | 0.0642593 | 0.00218449 |
| 5.55 | 5.57824 | 0.00508905 |
| 0.0201 | 0.020333 | 0.0115899 |
| 0.577 | 0.577234 | 0.000405281 |
| 2.15 | 2.1874 | 0.0173975 |

# References

[1] S. Bertoluzza. *Adaptive wavelet collocation method for the solution of Burgers equation.* Transport Theory and Statistical Physics, Vol. 25, (1996).

[2] H. G. Bock. *Numerical Treatment of Inverse Problems in Chemical Reaction Systems.* Springer Series in Chem. Phys., Vol. 18, (1981).

[3] H. G. Bock. *Randwertproblemmethoden zur Parameteridentifizierung in Systemen nichtlinearer Differentialgleichungen.* Bonner Mathematische Schriften Nr. 183, (1987).

[4] T. S. Carlson, J. Dockery, J. Lund. *A Sinc-Collocation Method for Initial Value Problems.* Mathematics and Computation, Vol. 66, No. 217, (1997).

[5] D. L. Donoho. *Interpolating Wavelet Transforms.* Tech. Rept. 408. Department of Statistics, Stanford University, Stanford, (1992).

[6] E. Hairer, G. Wanner. *Vol. 1 : Nonstiff Problems.* Springer, 2. Ed., (1993).

[7] E. Hairer, G. Wanner. *Vol. 2 : Stiff and Differential-Algebraic Problems.* Springer, 2. Ed., (1996).

[8] S. Kameswaran, L. T. Biegler. *Simultaneous Dynamic Optimization Strategies: Recent Advances and Challenges.* Computers and Chemical Engineering, Vol. 30, (2006).

[9] T. Lohmann, H. G. Bock, J. P. Schlöder. *Numerical Methods for Parameter Estimation and Optimal Experiment Design in Chemical Reaction Systems.* Ind. Eng. Chem. Res., Vol. 31, (1992).

[10] L. Qian. *On the Regularized Whittaker-Koltel'nikov-Shannon Sampling Theorem.* Proceedings of the Amarican Mathematical Society, Vol. 131, No. 4, (2002).

[11] M. Schuchmann. *Approximation and Collocation with Wavelets. Approximations and Numerical Solving of ODEs, PDEs and IEs*. Osnabrück, DAV, (2012).

[12] G. Strang. *Wavelets and Dilation Equations: A Brief Introduction.* SIAM Review Vol. 31, No. 4, (1989).

[13] M. Unser, T. Blu. *Comparison of Wavelets from the Point of View of their Approximation Error.* Proc. of SPIE Vol. 3458, Wavelet Applications in Signal and Image Processing, (1998).

[14] M. Unser. *Vanishing moments and the approximation power of wavelet expansions.* Proceedings of the 1996 IEEE International Conference on Image Processing, (1996).

[15] R. Vuduc. *A Wavelet Collocation Method for Solving PDEs*. J. Comp. Phys., Vol. 165 (2000).

## Notes on Nonparametric Regression with Wavelets

M. Schuchmann and M. Rasguljajew from the Darmstadt University of Applied Sciences

### Abstract

In this article we describe a nonparametric regression using a wavelet basis. There exist different approaches for a regression based on wavelets. For the regression which we use in our article we must calculate coefficients over an integral but for further regressions with the same number of points we can use the same coefficients. We describe the regression for points in $R^2$ and $R^3$ and we use a help function, which is constant on an area around the points in contrast to other approaches where the regression function is shifted.

### Regression for Functions $f: R \rightarrow R$

We have got $n$ points $(x_1,y_1)$, $(x_2,y_2)$, ...., $(x_n,y_n)$ for example from measurements. We assume, that there exists a causal relationship between the $x_i$ and the $y_i$, like

(1) $$y_i = f(x_i) + e_i \ .$$

The $e_i$ represents the error, for example the measurements error. Like in the classical statistics $y_i$ can be a realisation of a random variable, so that we have the theoretical model

$$Y_i = f(x_i) + E_i \ .$$

$e_i$ is a random variable with mean 0 and variance $\sigma^2$. With additional assumptions we can assume, that the $e_i$ are independent identically $N(0,\sigma^2)$ distributed, but that's not necessary for our method. The function $f$ is often unknown in praxis, but in the classical regression analysis the type is known. Here even the type of the function $f$ can be unknown.

If we apply a continuous approximation, where we knew the function $f$, we get an orthogonal projection form $f$ on $V_j$ with

$$f_j(x) = \sum_k f_k^j \phi_{j,k}(x),$$

with $$f_k^j = \int_{-\infty}^{\infty} \phi_{j,k}(x) f(x) dx \ .$$

For easier notation we assume, that the variable $x$ has values out of the interval [0, 1] and $x_i = i/n$, with $i = 1,2,..,n$. With the resolution $j$ we could adjust, how many or how small the details are, that we want take in account. Because $f$ is now unknown, we define a function $\tilde{f}$ which has around the point $x = x_i$ the constant function value $y_i$, so

$$\widetilde{f}(x) = y_i \text{ for } \frac{i-1}{n} + \frac{1}{2n} \le x < \frac{i}{n} + \frac{1}{2n}.$$

So we define:

$$\widetilde{f}(x) = \begin{cases} y_i & ; \frac{2(i-1)+1}{2n} \le x < \frac{2i+1}{2n} \\ 0 & ; else \end{cases}$$

We define the help function $\widetilde{f}$ so, that the function is constant around the measurement point $x_i$. There exists definitions, where the help function is constant from on $[x_i, x_{i+1}]$, but here the graph of regression function is shifted to the right.

For easier notation we assume that the scaling function is real-valued. Now we calculate a best approximation $\widetilde{f}_j$ of $\widetilde{f}$ in $V_j$ and we use this function as a regression function of the points $(x_i, y_i)$. So we get an approximation of the coeffitions $f_k^j$ with:

$$\widetilde{f}_k^j = \int_{-\infty}^{\infty} \phi_{j,k}(s)\widetilde{f}(s)ds = \sum_{i=1}^{n} \int_{\frac{2(i-1)+1}{2n}}^{\frac{2i+1}{2n}} \phi_{j,k}(s)\widetilde{f}(s)ds = \sum_{i=1}^{n} y_i \int_{\frac{2(i-1)+1}{n}}^{\frac{2i+1}{2n}} \phi_{j,k}(s)ds$$

So we get $\widetilde{f}_j$:

$$\widetilde{f}_j(x) = \sum_{k=-\infty}^{\infty} \widetilde{f}_k^j \cdot \phi_{j,k}(x) = \sum_{k=-\infty}^{\infty} \sum_{i=1}^{n} y_i \underbrace{\int_{\frac{2(i-1)+1}{2n}}^{\frac{2i+1}{2n}} \phi_{j,k}(s)ds}_{:=a^j_{i,k}} \cdot \phi_{j,k}(x)$$

$$= \sum_{i=1}^{n} y_i \sum_{k=-\infty}^{\infty} a^j_{i,k} \cdot \phi_{j,k}(x)$$

If we use the same number of measurement points $n$, we don't need do calculate the integral above twice, we can use the same coefficients $a^j_{i,k}$.

In practice we don't need the whole summation area $\mathbb{Z}$ for the index $k$, because we have (here [0, 1]) a compact interval and either the scaling function $\phi$ has compact support or it will vanish for big or small arguments.

### Example 1:

We simulated measurement points like in formula (1) and used a normal distributed error with mean $\mu = 0$ and standard deviation $\sigma = 0.01$. We used the function $f$:

$$f(t) = \begin{cases} \sin(2\pi \cdot t) & \text{if } 0 \le t < 1/2 \\ \sin(4\pi \cdot t) & \text{if } 1/2 \le t < 1 \\ 0 & \text{else} \end{cases}$$

We set $n = 20$. In the graph we see, that $f$ is not differentiable an the point $x = \frac{1}{2}$:

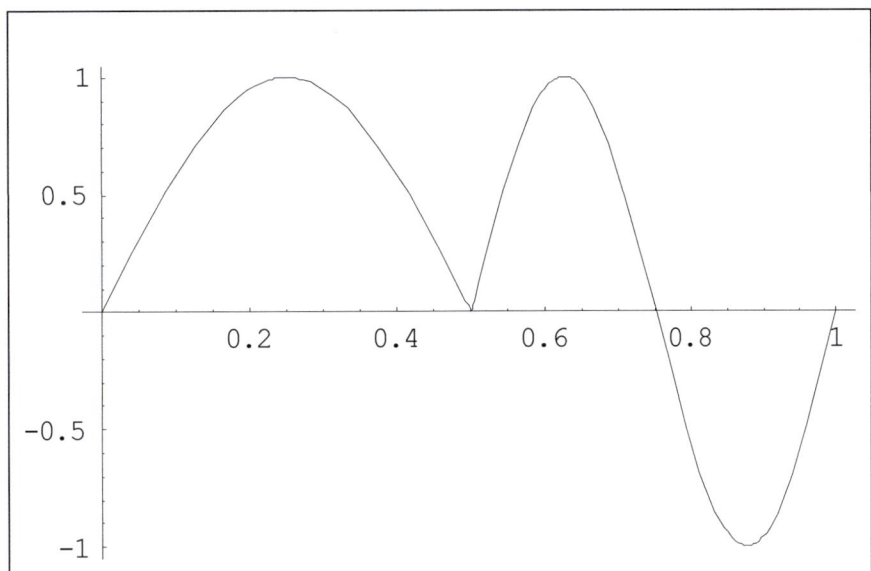

Figure 1: Graph of $f$

Here is the plot of the points:

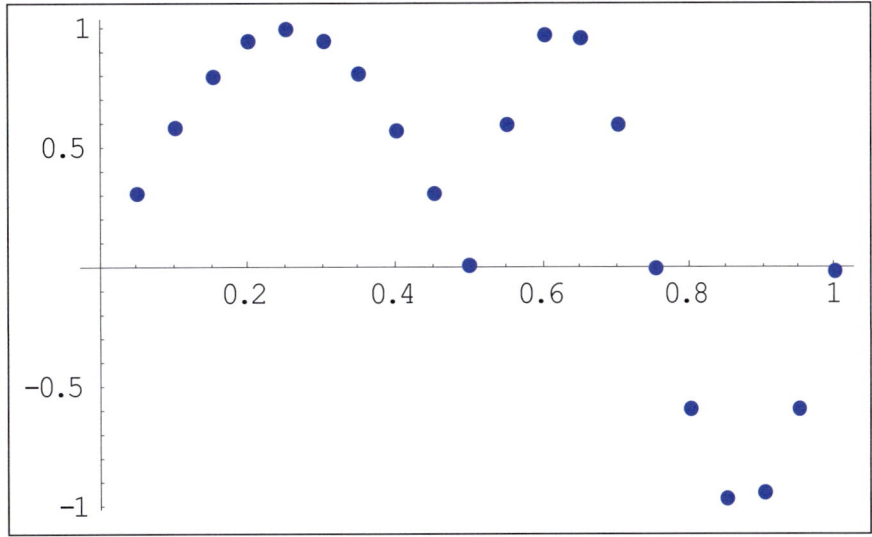

Figure 2: Simulated measurement points

Using the Haar wavelet und setting $j = 4$, $k = -16,\ldots,16$, we get the following regression function:

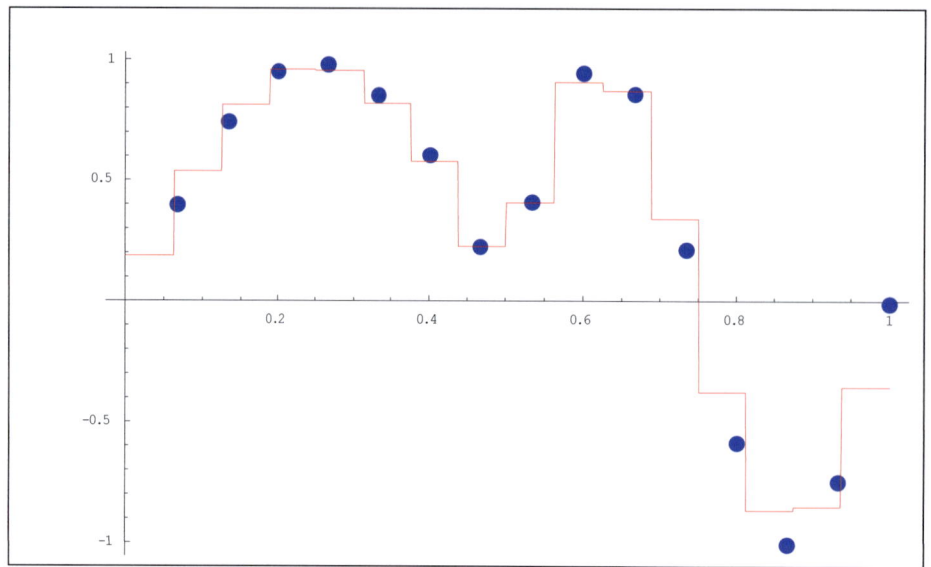
Figure 3: Graph of the regression function using the Haar wavelet

With the Shannon wavelet we get a useful regression function:

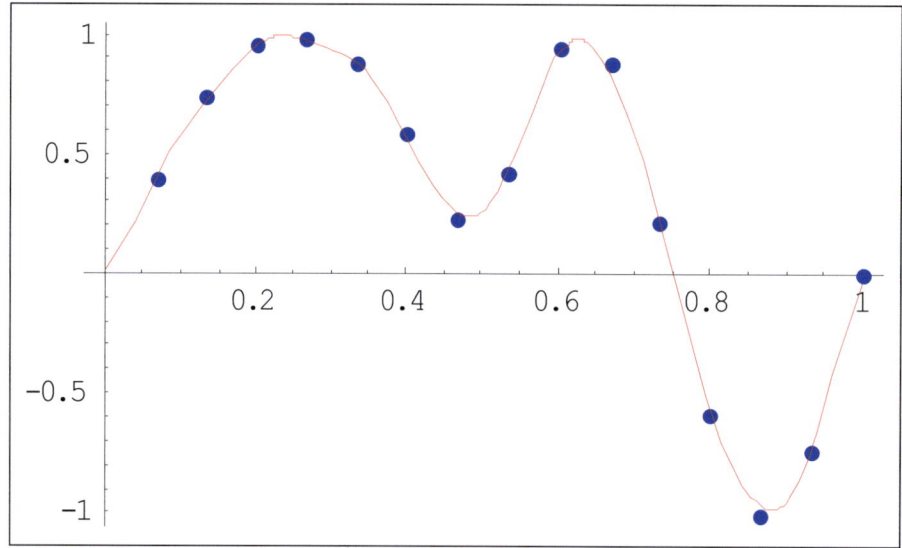
Figure 4: Graph of the regression function using the Shannon wavelet

### Remarks 1:
1) The nonparametric regression with the Shannon wavelet can be applied under
http://www.statistikcloud.at/htdocs-eng/Wavelet-Regression-v3 (for small examples).

2) A method for discrete approximation, which causes less effort, is the direct application of the Least Squares Method. So we would solve

$$\min Q(a)$$
$$\text{with } Q(a) = \sum_{i=1}^{n}(y_i - f_j(t_i))^2$$

and

$$f_j(t) = \sum_{k=k_{min}}^{k_{max}} a_k \phi_{j,k}(t).$$

This is a quadratic problem, because $f_j$ is linear in $a$.

# Regression for Functions $f: R^2 \to R$

We assume that we got points $(s_i, t_l, y_{i,l})$ and

(2)    $y_{i,l} = f(s_i, t_l) + e_{i,l}$ with $i = 1,2,...,n_1$, $l = 1,2,...,n_2$ and $(s_i, t_l) \in G$

with

$$f: R^2 \to R.$$

Here are $e_{i,l}$ the errors, which are for example realisations of independent identically $N(0, \sigma^2)$ random variables $E_{i,l}$. We assume $s_i \neq s_j$ and $t_i \neq t_j$ for $i \neq j$.

Now we choose a partition $(Z_{i,l})$ from $G$:

(1)    $\bigcup_{i,l} Z_{i,l} = G$ and $Z_{i,l} \cap Z_{i',l'} = \{\}$ with $i \neq i'$ and $l \neq l'$.

(2)    $(s_i, t_l) \in Z_{i,l}$

We construct a function $\tilde{f}$, which has on the area $Z_{i,l}$ the constant function value $y_{i,l}$:

$$\tilde{f}(s,t) = y_{i,l} \text{ if } (s,t) \in Z_{i,l},$$
$$\tilde{f}(s,t) = 0 \text{ else.}$$

So $\tilde{f}$ looks like:

$$\tilde{f}(s,t) = \sum_{i=1}^{n_1} \sum_{l=1}^{n_2} y_{i,l} \cdot 1_{Z_{i,l}}(s,t),$$

with $1_A(s,t) = 1$ if $(s,t) \in A$ and $1_A(s,t) = 0$ else.

The regression function $\tilde{f}_j$ we get analogous to the two dimensional case:

$$\tilde{f}_j(s,t) = 2^j \sum_{k_1,k_2} \tilde{f}^j_{k_1,k_2} \cdot \phi(2^j s - k_1, 2^j t - k_2),$$

with

$$\tilde{f}^j_{k_1,k_2} = 2^j \cdot \int_{-\infty}^{\infty}\int_{-\infty}^{\infty} \tilde{f}(s,t) \cdot \phi(2^j s - k_1, 2^j t - k_2) \, ds\, dt$$

$$= 2^j \cdot \sum_{i=1}^{n_1} \sum_{l=1}^{n_2} y_{i,l} \cdot \underbrace{\iint_{Z_{i,l}} \phi(2^j s - k_1, 2^j t - k_2) ds dt}_{:= a^j_{i,k_1,k_2}}.$$

Using the same partition ($Z_{i,l}$), we need to evaluate the integral above only once if we save the coefficients $a^j{}_{i,k_1,k_2}$.

If the support of $\phi$ is compact, we must only take the following $k_1, k_2 \in Z$ in account:

$$\{(k_1, k_2) \mid supp\, \phi(2^j s - k_1, 2^j t - k_2) \cap G \neq \{\}\}.$$

If the support of $\phi$ is not compact, we use only the following $k_1, k_2 \in Z$ in the summation above:

$$\{(k_1, k_2) \mid |\phi(2^j s - k_1, 2^j t - k_2)| > \varepsilon \text{ for } (s,t) \in G\}$$

With a useful $\varepsilon > 0$. In many examples we saw, that the method is relatively insensitive and even with scaling functions without compact support we need not many basis coefficients for a good regression. Using an equidistant grid, we get the $s_i$ and $t_l$ with

and
$$s_i = s_1 + (i-1)h_s, \text{ for } 1 \leq i \leq n_1$$

$$t_l = t_1 + (l-1)h_t, \text{ for } 1 \leq l \leq n_2,$$

with

$$h_s = \frac{s_{n_1} - s_1}{n_1 - 1} \text{ and } h_t = \frac{t_{n_2} - t_1}{n_2 - 1}.$$

Here you see the grid an $Z_{i,l}$ in a graph:

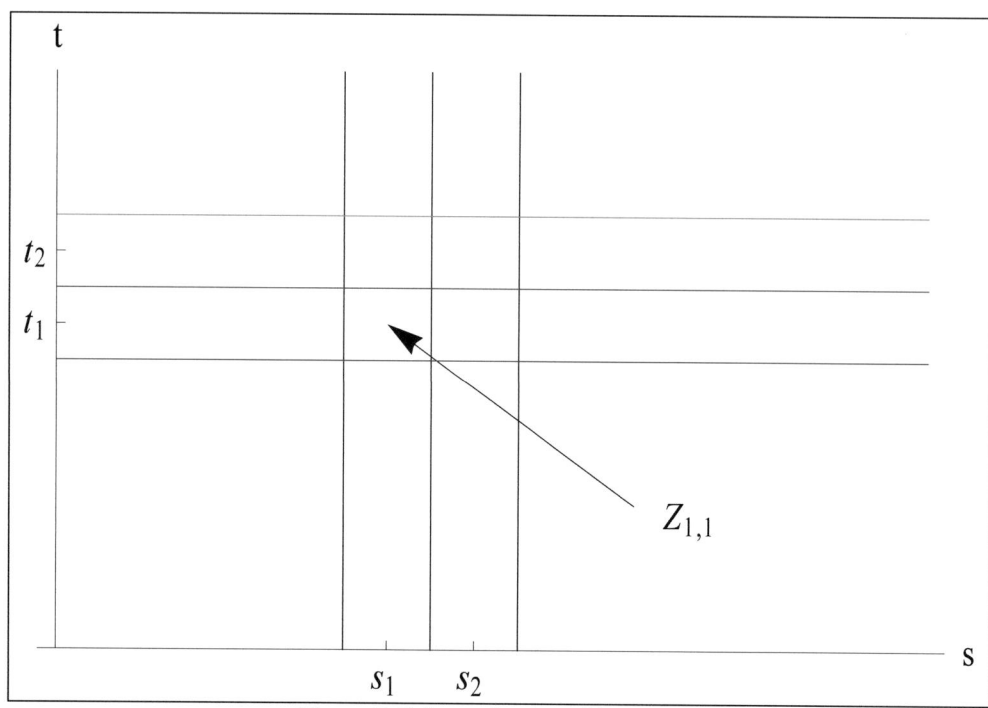

Figure 5: Scheme for the grid

Now the $Z_{i,l}$ are:

$$Z_{i,l} = [s_i - h_s/2, s_i + h_s/2) \times [t_l - h_t/2, t_l + h_t/2)$$

The formula for the coefficients $\tilde{f}^j_{k_1,k_2}$ has then the following form:

$$\tilde{f}^j_{k_1,k_2} = 2^j \cdot \sum_{i=1}^{n_1} \sum_{l=1}^{n_2} y_{i,l} \cdot \int_{t_l-h_t/2}^{t_l+h_t/2} \int_{s_i-h_s/2}^{s_i+h_s/2} \phi(2^j s - k_1, 2^j t - k_2)\, ds\, dt$$

**Example 2:**
We generated points by simulation measurements (formula (2)) and used the function

$$f(s,t) = e^{-s^2-t^2}.$$

We generated 100 function values over the area $[-3, 3]^2$ at equidistant points. The error was chosen normal distributed with mean 0 and the standard deviation 0.001.

We set:
$$n_1 = n_2 = 10;$$
$$s_1 = t_1 = -3;$$
$$s_{n_1} = t_{n_2} = 3.$$

Here you see the points $(s_i, t_l, y_{i,l})$ together with the graph of f:

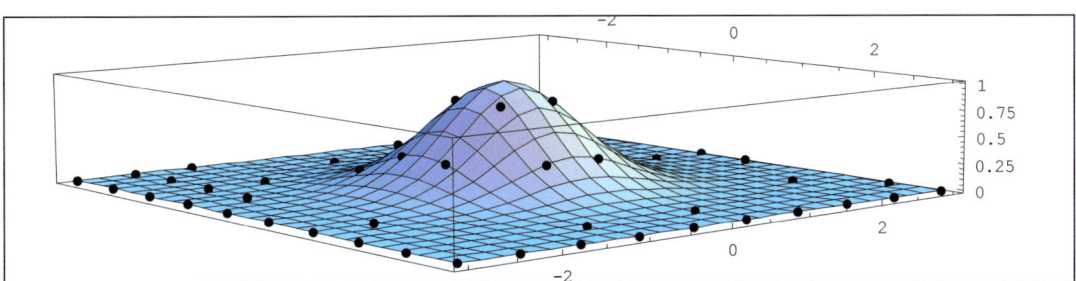

Figure 6: Graph from f and the regression points

We use the Daubechies wavelet of order 7. Here is the graph of the one dimensional scaling function:

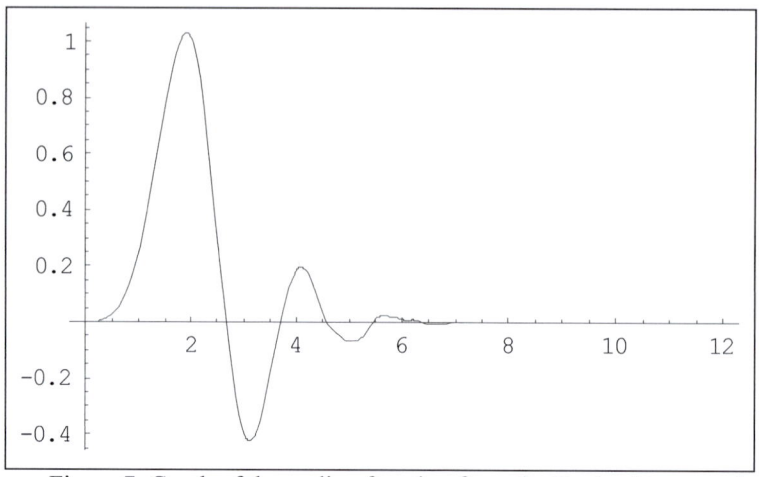

Figure 7: Graph of the scaling function from the Daubechies wavelet

And here is the graph of the scaling function with two arguments:

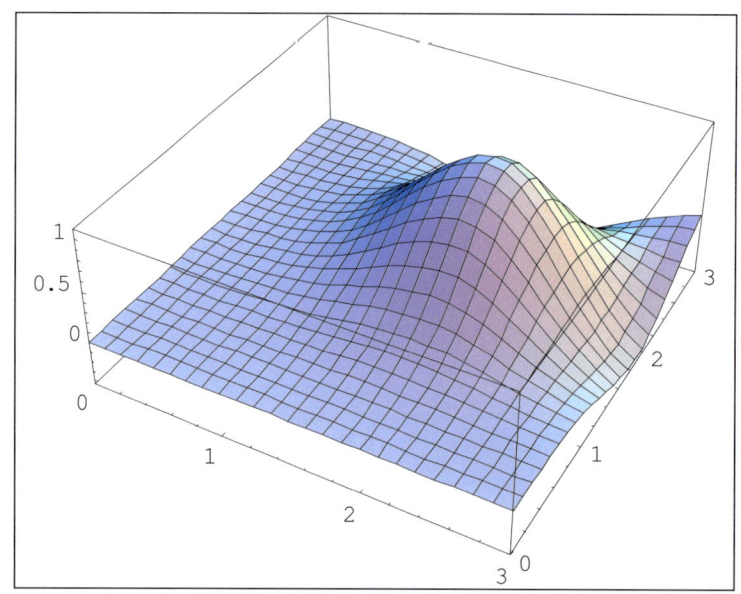

Figure 8: Graph of the scaling function with two arguments from the Daubechies wavelet

Here is the graph of the approximation function at resolution *j = 0*:

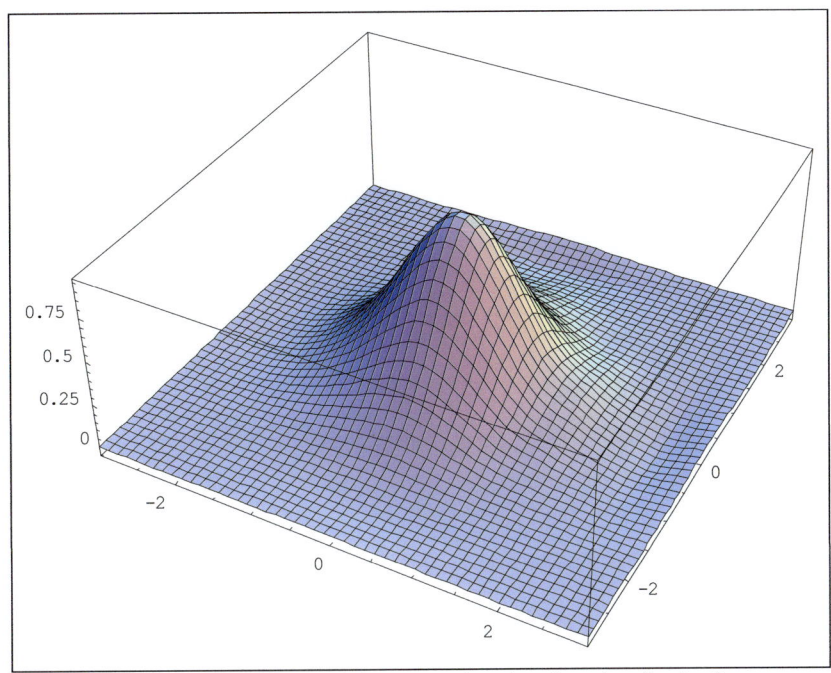
Figure 9: Graph of the approximation function for $j = 0$

The graph of the approximation function at resolution $j = 1$:

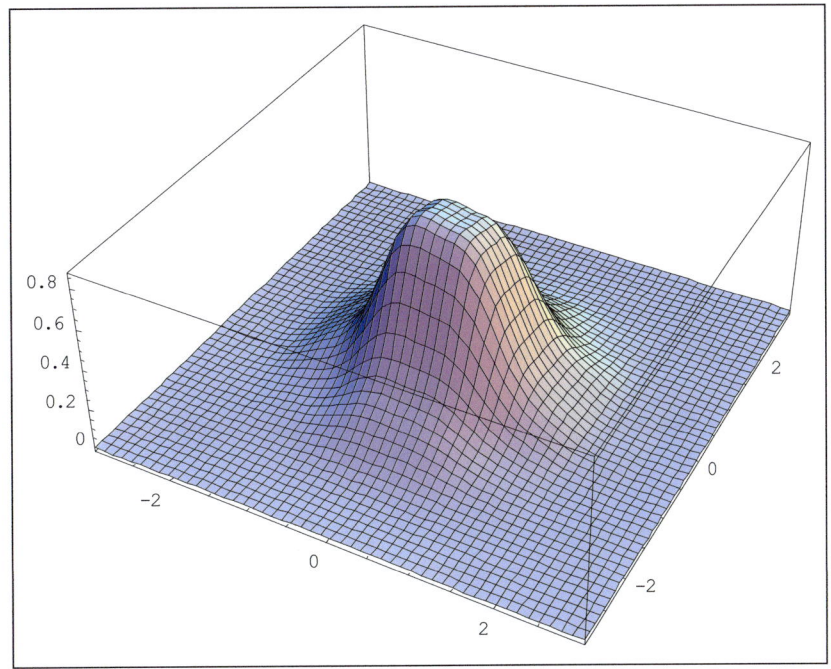
Figure 10: Graph of the approximation function for $j = 1$

The graph of the details $d_0$ ($= f_1 - f_0$):

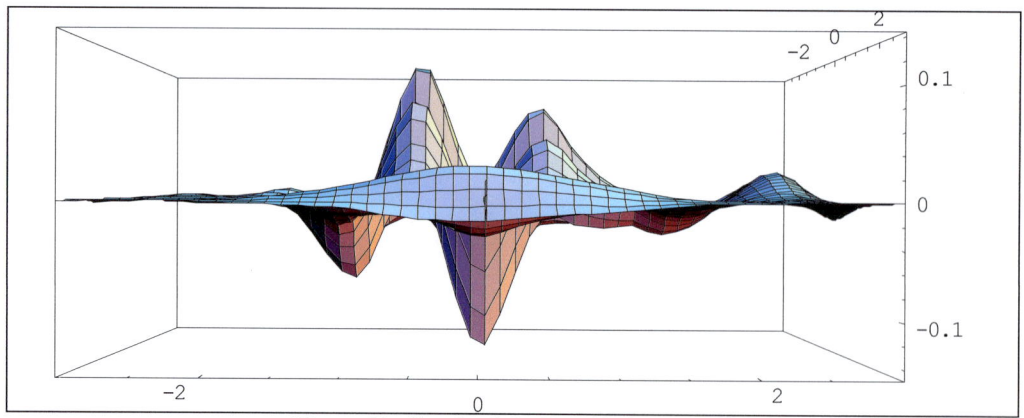

Figure 11: Graph of the Details for $j = 0$

The indices $k_1$ and $k_2$ were chosen from $-11 \cdot 2^j$ to $3 \cdot 2^j$. We calculate the mean quadratic error at both resolutions:

$$\frac{1}{n_1 \cdot n_2} \sum_i^{n_1} \sum_l^{n_2} (y_{i,l} - f_0(s_i, t_l))^2 = 0.00043196 ,$$

$$\frac{1}{n_1 \cdot n_2} \sum_i^{n_1} \sum_l^{n_2} (y_{i,l} - f_1(s_i, t_l))^2 = 0.000261559 .$$

## References

[1] R. T. Ogden. *Essential Wavelets for Statistical Applications and Data Analysis.* Boston, Berlin, Basel: Birkhäuser, (1997).

[2] M. Schuchmann. *Approximation and Collocation with Wavelets. Approximations and Numerical Solving of ODEs, PDEs and IEs.* Osnabrück, DAV, (2012).

# Extrapolation and Approximation with a Wavelet Collocation Method for ODEs

M. Schuchmann and M. Rasguljajew from the Darmstadt University of Applied Sciences

### Abstract

In this article we use a wavelet collocation method for an approximation of the solution of an ODE. We show that an approximation with the Shannon wavelet leads to better approximations than a Daubechies wavelet and we even can use the approximation for an extrapolation.

### Introduction

We use the same Method as in "An Approximation on a Compact Interval Calculated with a Wavelet Collocation Method can Lead to Much Better Results than other Methods" described.

In the wavelet theory a scaling function $\phi$ is used, which belongs to a MSA (multi scale analysis). From the MSA we know, that we can construct an orthonormal basis of a closed subspace $V_j$, where $V_j$ belongs to a the sequence of subspaces with the following property:

$$\ldots \subset V_{-1} \subset V_0 \subset V_1 \subset \ldots \subset L^2(R),$$

$\{\phi_{j,k}(t)\}_{k \in Z}$ is an orthonormal basis of $V_j$ with $\phi_{j,k}(t) = 2^{j/2}\phi(2^j t - k)$.

We use the following approximation function

$$y_j(t) := \sum_{k=k_{min}}^{k_{max}} c_k \cdot \phi_{j,k}(t) \quad \text{, with } \phi \in C^1(R).$$

$k_{max}$ and $k_{min}$ depend on the approximation interval $[t_0, t_{end}]$ (see [7]).

Now we can approximate the solution of an initial value problem $y' = f(y,t)$ and $y(t_0) = y_0$ by minimization of the following function

$$(1) \quad Q(c) = \sum_{i=1}^{m} \left\| y_j'(t_i) - f(y_j(t_i), t_i) \right\|_2^2 + \left\| y_j(t_0) - y_0 \right\|_2^2.$$

We apply the describe method in an example:

# Applying the Method and Assessing an Approximation

**Example 1:**
We want to approximate the solution of

$$y' = -t\,y,$$
$$y(0) = 1$$

on the interval $I = [t_0, t_{end}] = [0, 3]$.

We know minimize $Q$ for $k_{max} = -k_{min} = 15, 20, 15$ with $j = 0, 1, 2$ and $h = (t_{end} - t_0)/(4k_{max}) = 3/(4k_{max})$.

We calculate the mean squared error

$$(2) \qquad mse = \frac{1}{101}\sum_{i=0}^{100}(y(t_0 + i \cdot h_0) - y_j(t_0 + i \cdot h_0))^2$$

with $h_0 = (t_{end} - t_0)/100 = 3/100$ and $Q_{min} = \min Q(c) = Q(\hat{c})$. $sse$ is the sum of squared errors with $sse = mse \cdot 101$.

Here we see the results of the graphs from $y_j$ and $y$.

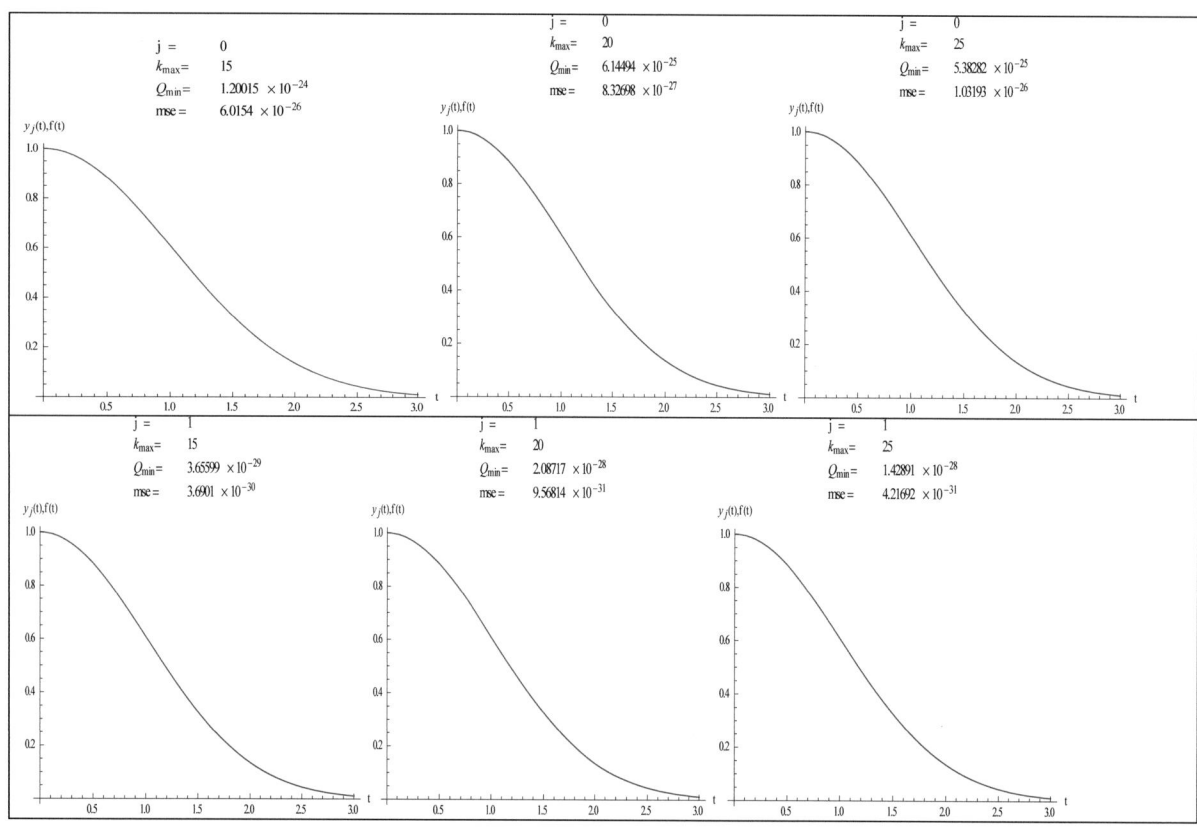

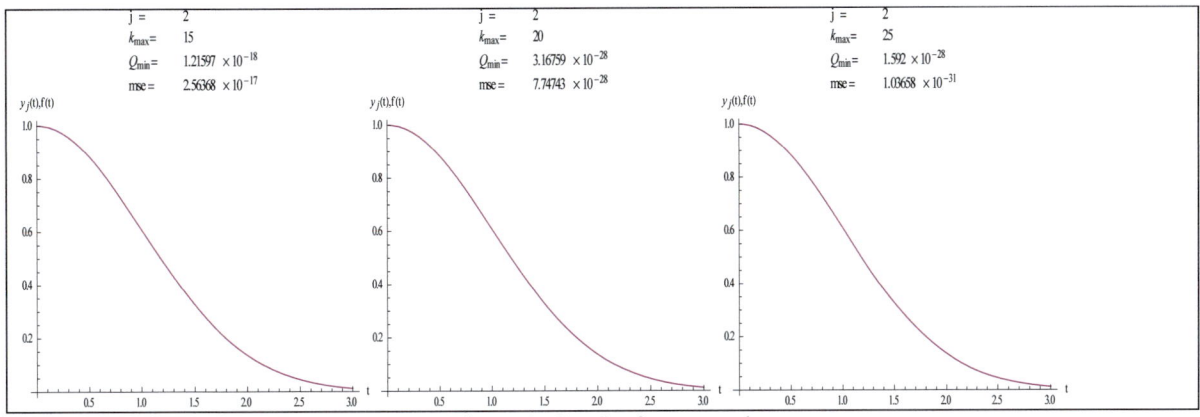

Figure 1. Graphs from $y_j$ and $y$

As seen above, there is a correlation between $Q_{min}$ and *mse*.

We now apply a linear regression on the points *(-ln($Q_{min}$), -ln(mse))* (here we have: (55.0254, 53.4888), (55.749, 55.4352), (55.8814, 55.2207), (65.4786, 63.1568), (63.7366, 64.5066), (64.1155, 65.3259), (41.251, 33.5874), (63.3194, 57.8099), (64.0074, 66.7291))

for the different $j$ und $k_{max}$:

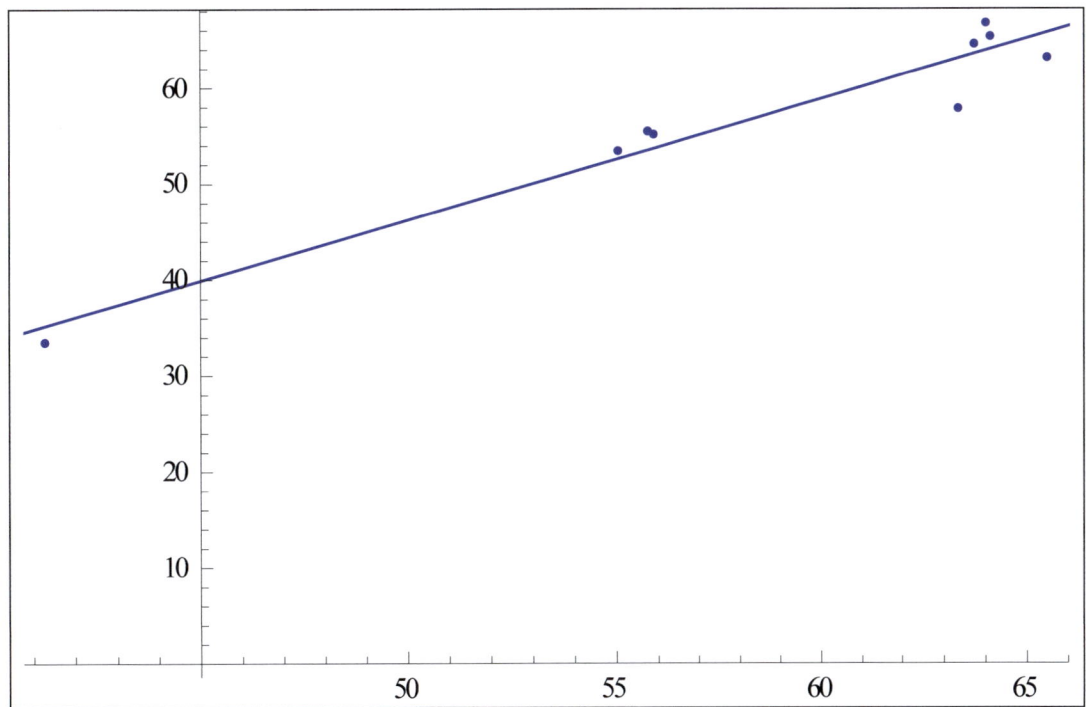

Figure 2. Linear Regression on the points *(-ln($Q_{min}$), -ln(mse))*

Here is the regression table with a $R^2$ of 0.934136.

|   | Estimate | SE | TStat | PValue |
|---|---|---|---|---|
| 1 | -16.7718 | 7.48689 | -2.24016 | 0.0600638 |
| x | 1.26041 | 0.126497 | 9.96392 | 0.0000219101 |

So we can see relativ good with $Q_{min}$, if an approximation is good. But it can occur that the residuals are very small at the collocation points but not between them. For that reason we later define $Q_a$ to detect this.

Now we see the graph of $d$ for $k_{max} = 25$ and $j = 2$ with

(3) $$d(t) = \left\| y_j{'}(t_i) - f(y_j(t_i), t_i) \right\|_2^2:$$

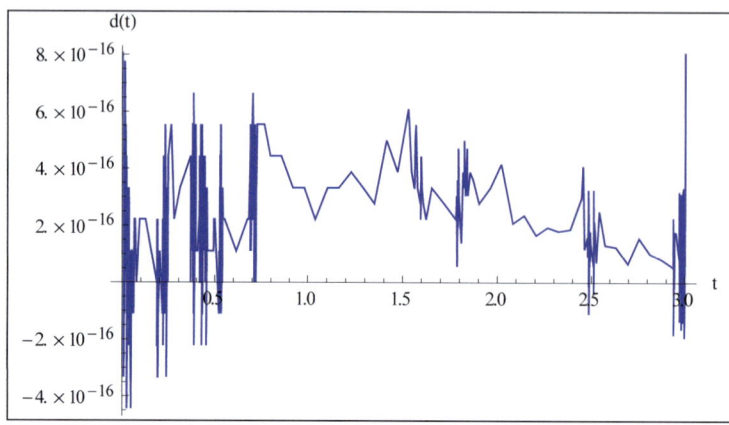

Figure 3. Graph of $d$ for $k_{max} = 25$ and $j = 2$

Here is the Graph of $(k, -ln(mse))$:

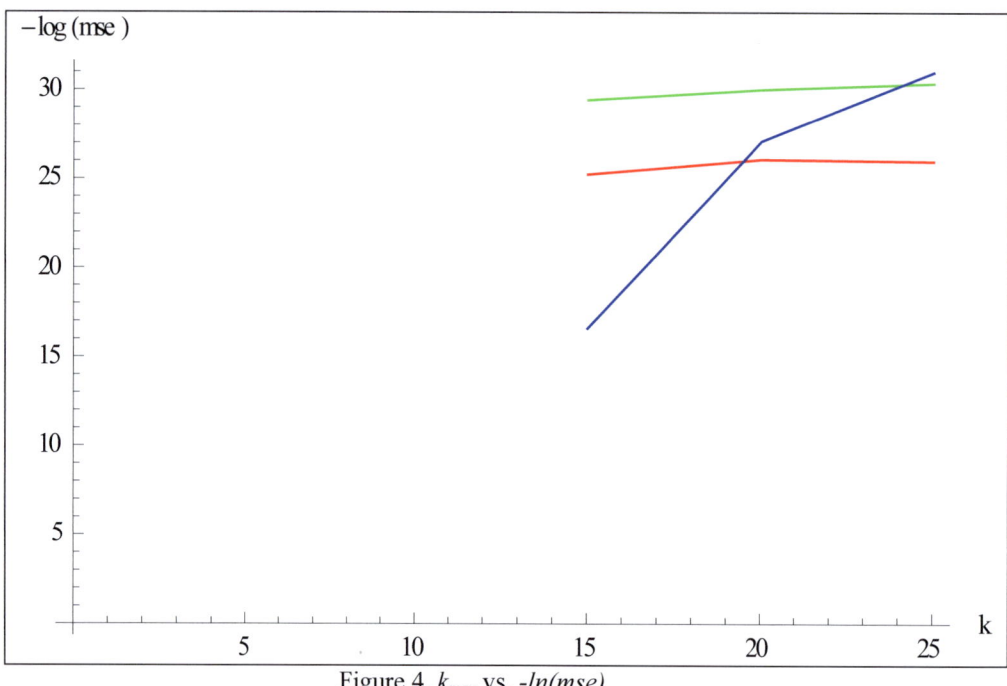

Figure 4. $k_{max}$ vs. $-ln(mse)$

In red we see the points for $j = 0$ connected with lines, in green for $j = 1$ and in blue for $j = 2$. Theoretically the curves must increase because for a greater $k_{max}$ the values of $Q_{min}$ must shrink.

We have seen in this example, that there is a correlation between $Q_{min}$ and $mse$. This relationship can - with an insufficient number of collocation points - no longer exist. Here, however, we can simply use another criterion $Q_a$ where we can simply use the calculated $\hat{c}$ from the minimization:

(4) $$Q_a(\hat{c}) = \sum_{i=1}^{m_a} \left\| y_j{'}(\tau_i) - f(y_j(\tau_i), \tau_i) \right\|_2^2 + \left\| y_j(t_0) - y_0 \right\|_2^2$$

with $\tau_i = t_0 + i \cdot h/a$. $m_a = a \cdot m$ and an integer $a > 1$. If we use a big $a$, we should weight $Q_a$ with $1/a$, but in different simulation we got with $a = 2$ good results:

(4a) $$\widetilde{Q}_a(\hat{c}) = 1/a \cdot \sum_{i=1}^{m_a} \left\| y_j{}'(\tau_i) - f(y_j(\tau_i), \tau_i) \right\|_2^2 + \left\| y_j(t_0) - y_0 \right\|_2^2$$

For a good approximation $Q_a$ should be small with any $a$. If h is too big, than $Q_a \gg Q_{min}$.

**Example 2:**
We now calculate the approximations in example 1 with another $h$, which is too big $h = (t_{end} - t_0)/(1/2k_{max}) = 6/k_{max}$. Here we see with the following graphs, that $Q_{min}$ can be small but *mse* and *sse* a relativ big (and so the approximation is not good). The worse approximation we can detect with $Q_2$ or $Q_4$.

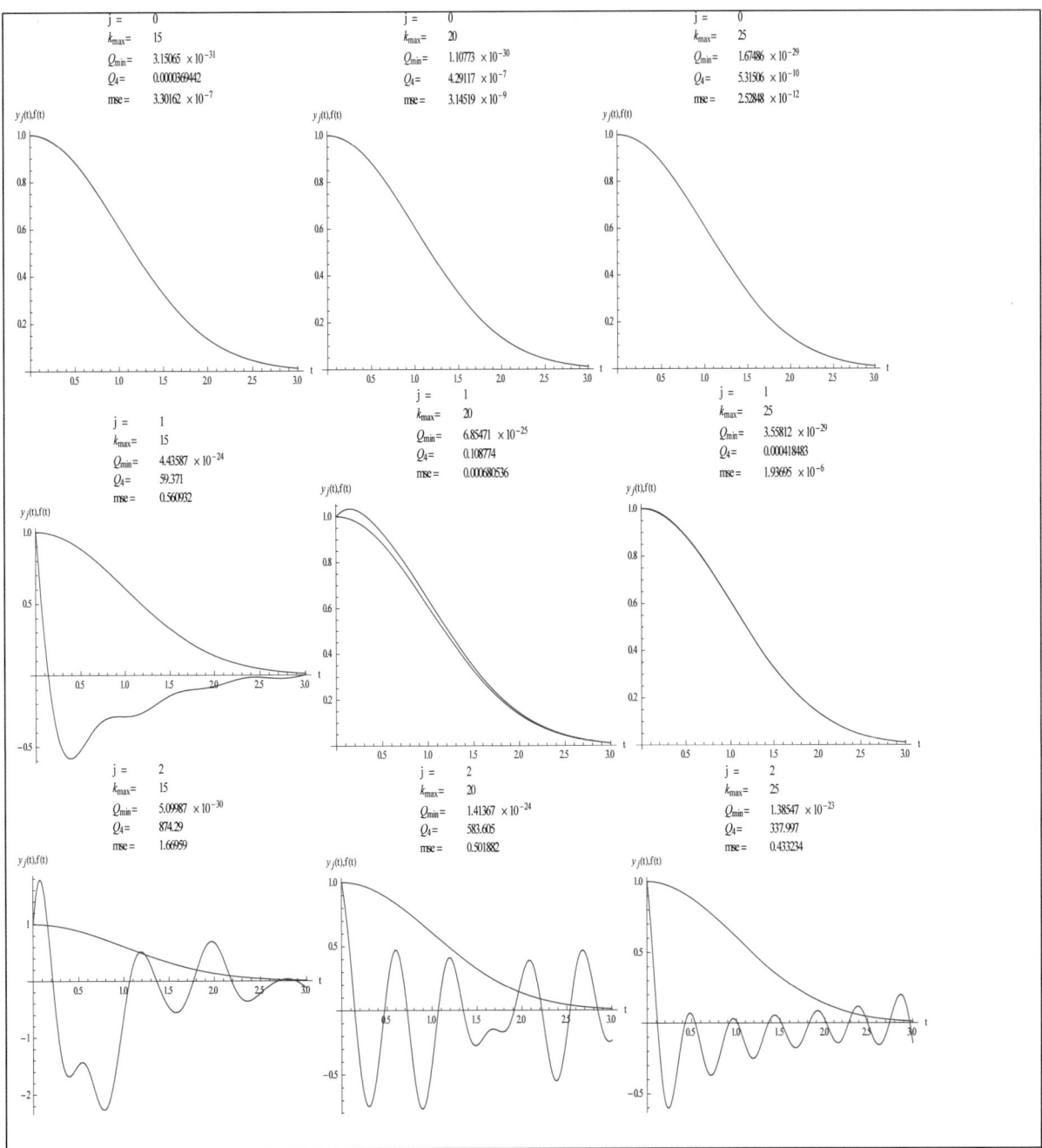

Figure 5. Graphs from $y_j$ and $y$ with a too big $h$

Now we see three linear regressions with the points $(-ln(Q_{min}), -ln(mse))$ the points $(-ln(Q_4), -ln(mse))$ and $(-ln(Q_2), -ln(mse))$. Here we can see, that the correlation between $Q_{min}$ and *mse* is because of the too big step size not so high.

**$Q_{min}$ vs. mse:**

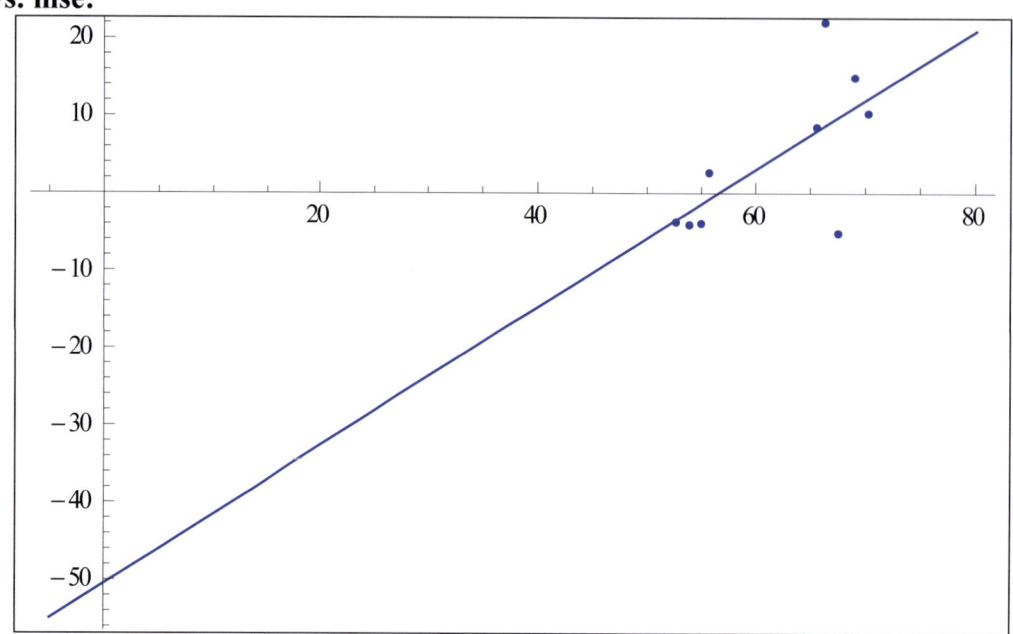

Figure 6. Linear Regression on the points $(-ln(Q_{min}), -ln(mse))$

Here is the regression table with a $R^2$ of 0.433152.

|   | Estimate | SE | TStat | PValue |
|---|----------|----|----|--------|
| 1 | -50.4893 | 23.9804 | -2.10544 | 0.0732805 |
| x | 0.893277 | 0.386234 | 2.31279 | 0.0539648 |

**$Q_4$ vs. mse:**

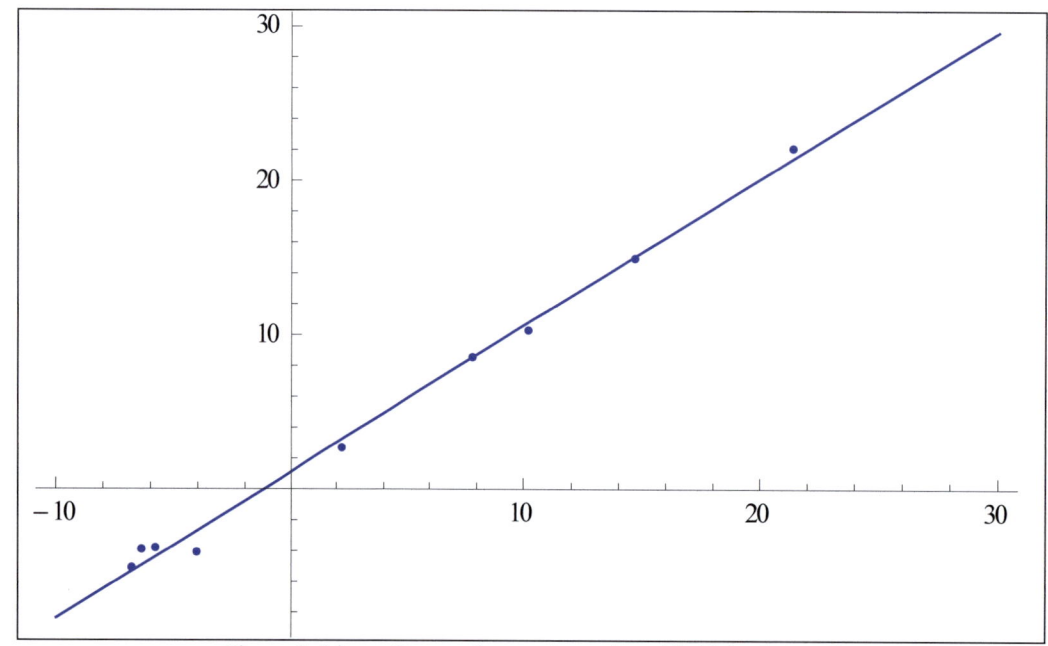

Figure 7. Linear Regression on the points $(-ln(Q_4), -ln(mse))$

Here is the regression table with a $R^2$ of 0.994715.

|   | Estimate | SE | TStat | PValue |
|---|----------|----|----|--------|
| 1 | 1.13039 | 0.272975 | 4.14099 | 0.00434376 |
| x | 0.950635 | 0.0261912 | 36.296 | $3.13045 \times 10^{-9}$ |

**$Q_2$ vs. mse:**

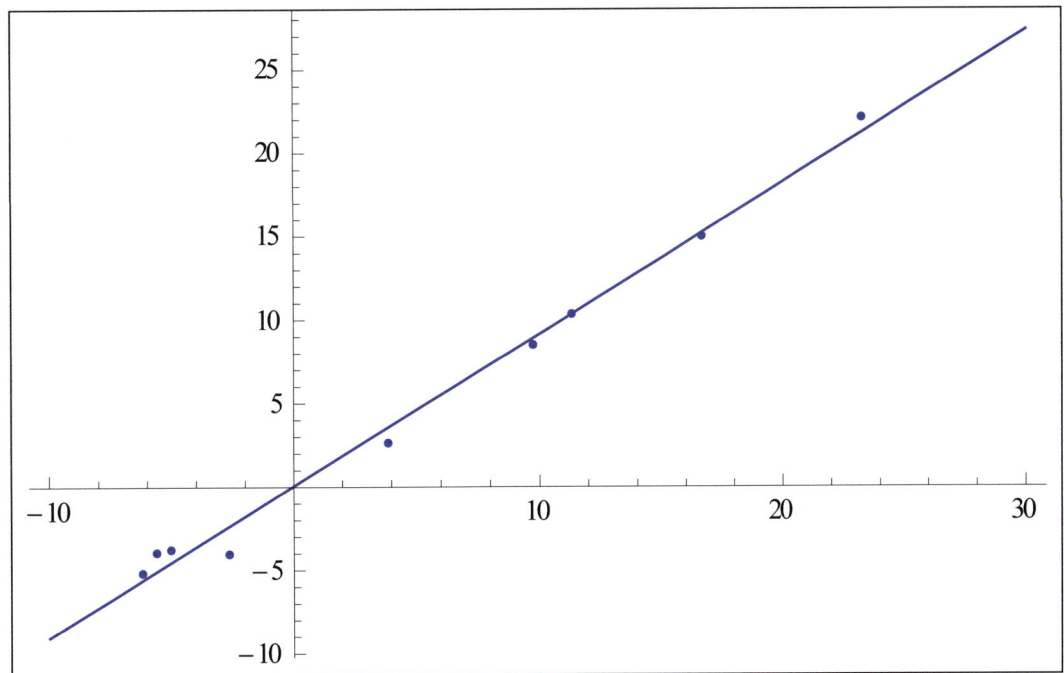

Figure 8. Linear Regression on the points $(-ln(Q_2), -ln(mse))$

Here is the regression table with a $R^2$ of 0.991411.

|   | Estimate | SE | TStat | PValue |
|---|----------|----|----|--------|
| 1 | 0.0730749 | 0.362907 | 0.20136 | 0.846143 |
| x | 0.90963 | 0.0320016 | 28.4245 | $1.71502 \times 10^{-8}$ |

So $Q_a$ is here a good criterion to detect a worse approximation.

For $j = 2$ and $k_{max} = 15$ the approximation was bad. Here was $h = 6/15 = 0.4$. We see with the graph of $d$ (see (3)) that the residuals are only small at the collocation points:

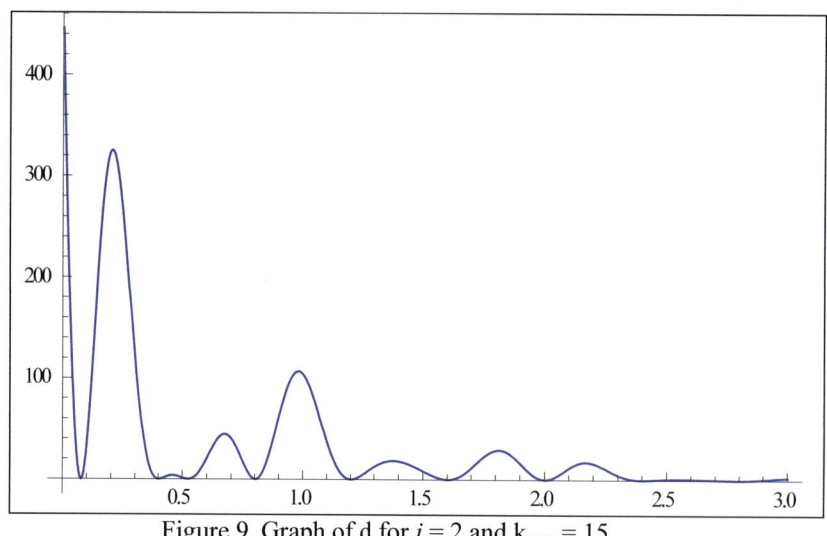
Figure 9. Graph of d for $j = 2$ and $k_{max} = 15$

Here we see how $d(t_i) = d(0.4i)$ is very small and between two collocation points $d$ has very big function values.

For $j = 0$ and $k_{max} = 20$ the approximation was good. Here $h = 6/20 = 0.3$. We see the graph for that case:

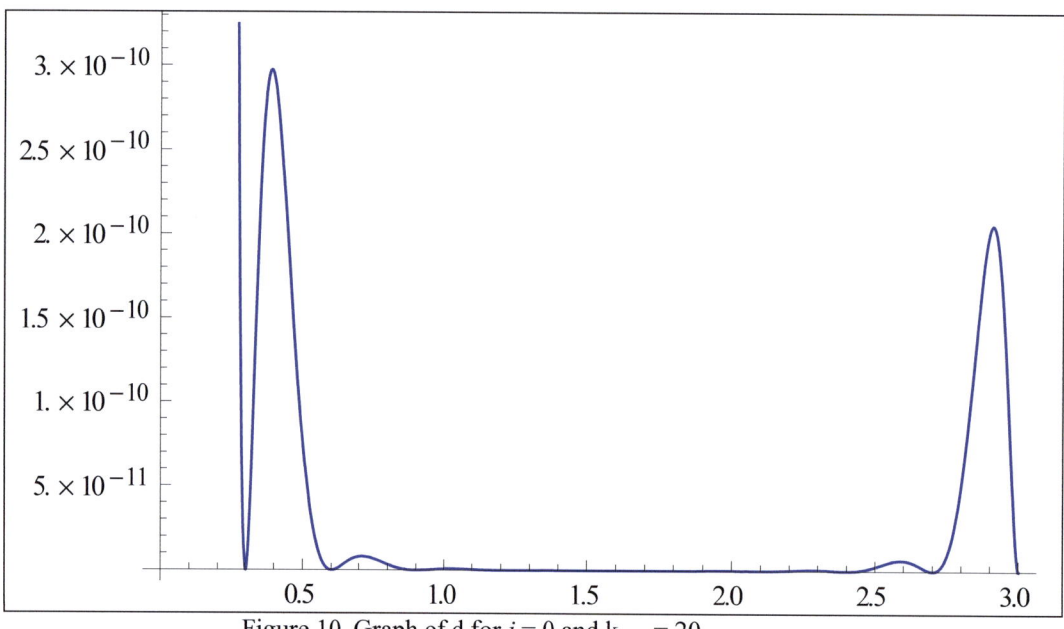
Figure 10. Graph of d for $j = 0$ and $k_{max} = 20$

Because we started with the collocation point $t_1$ in $Q$ we get a relative big value of $d$ at the point $t_0$.

Here we see the graph with the whole plot range:

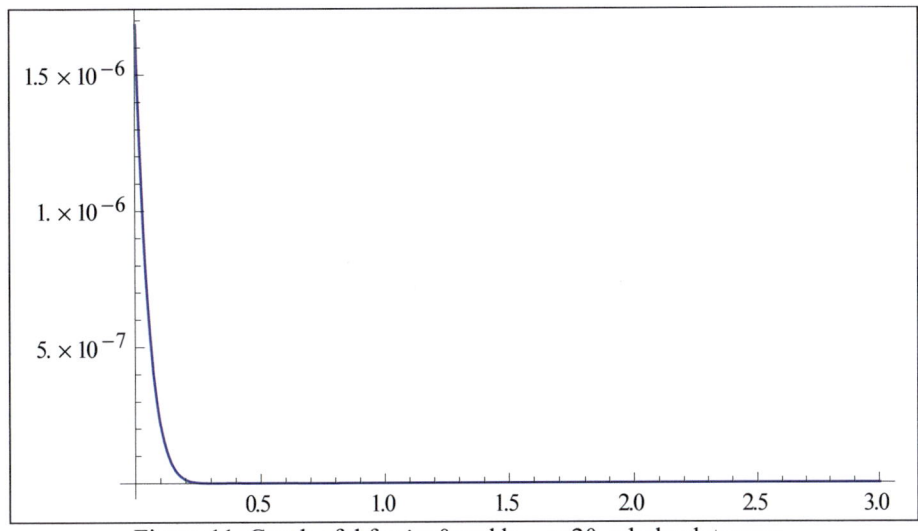

Figure 11. Graph of d for $j = 0$ and $k_{max} = 20$, whole plot range

## Using the Method for an Extrapolation

The approximation function can be even used for an extrapolation outside the approximation interval $[t_0, t_{end}]$.

We consider the approximations function $y_j$ for $j = 0$ and $k_{max} = 15$ from example 1 on the interval $[-2, 5]$:

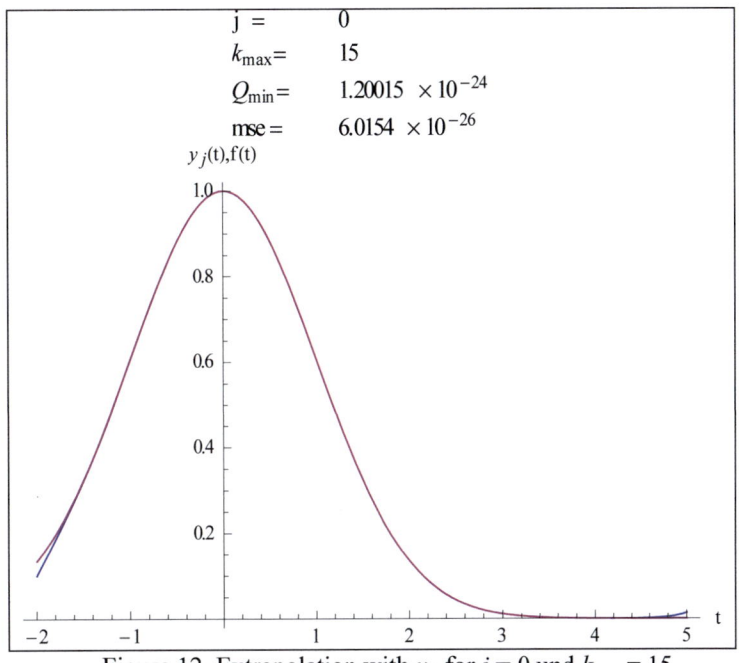

Figure 12. Extrapolation with $y_j$ for $j = 0$ und $k_{max} = 15$

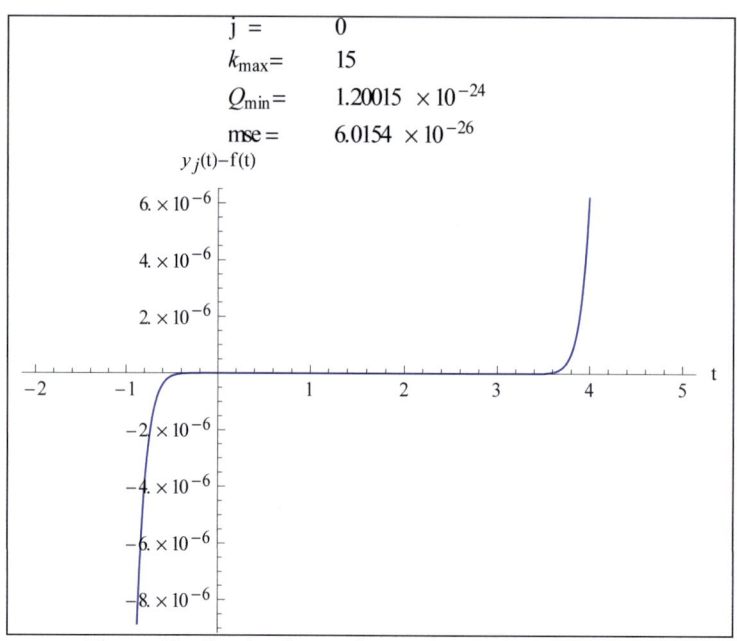

Figure 13. Graph of $y_j - y$ for $j = 0$ und $k_{max} = 15$.

If we use in example 1 the Intervall $I = [-1, 1]$ with $h = 2/m$ and $m = 2k_{max}$, then we get the following graph of the approximation function $y_j$ for $j = 0$ and $k_{max} = 15$:

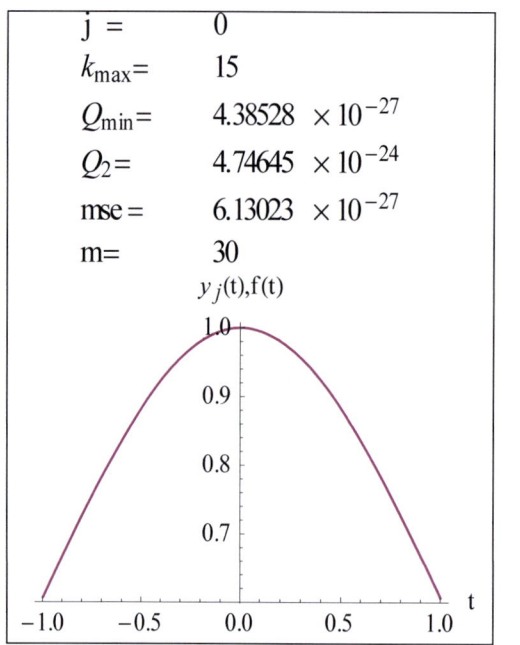

Figure 14. Graphs of $y_j$ and $y$ for $j = 0$ und $k_{max} = 15$

Here is the graph of the difference $y_j - y$:

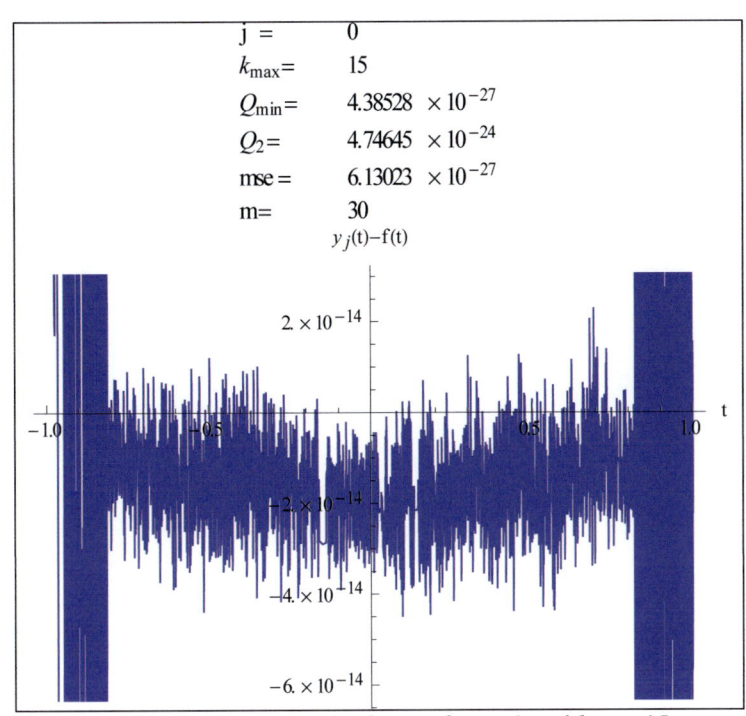

Figure 15. Graph of $y_j - y$ for $j = 0$ und $k_{max} = 15$

Now we consider this approximation function on a bigger interval [-2, 2]:

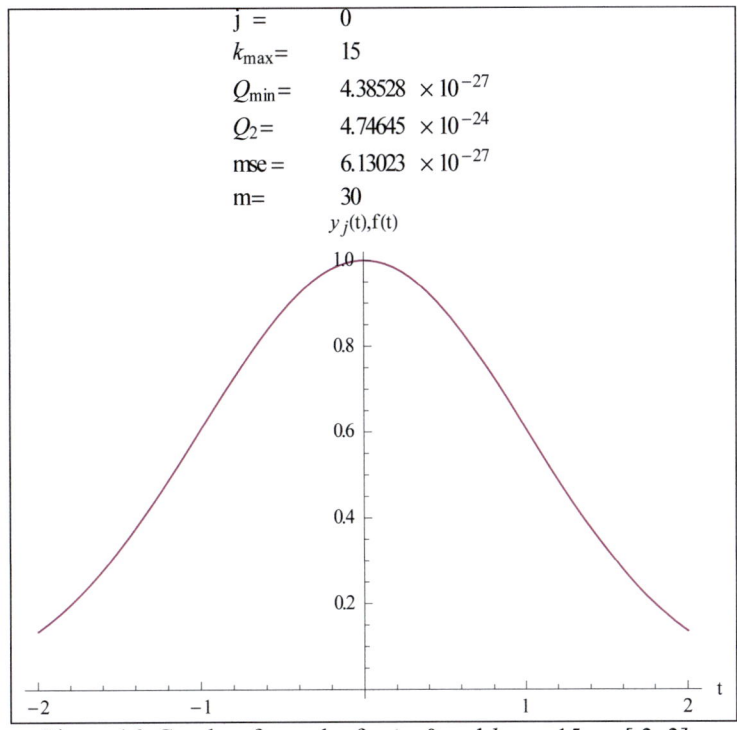

Figure 16. Graphs of $y_j$ and y for $j = 0$ und $k_{max} = 15$ on *[-2, 2]*

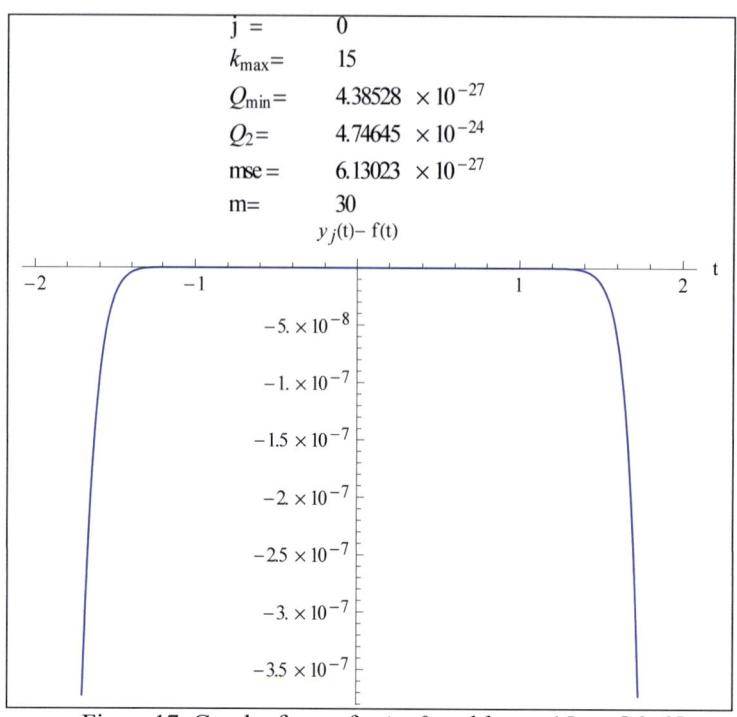

Figure 17. Graph of $y_j - y$ for $j = 0$ und $k_{max} = 15$ on $[-2, 2]$

Here is the approximation function on a three times bigger interval than the approximation interval:

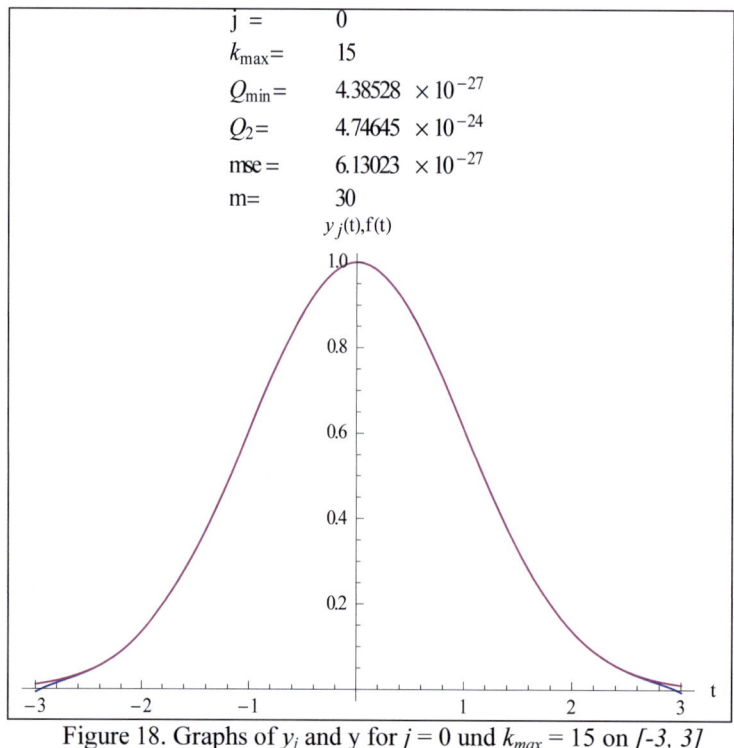

Figure 18. Graphs of $y_j$ and $y$ for $j = 0$ und $k_{max} = 15$ on $[-3, 3]$

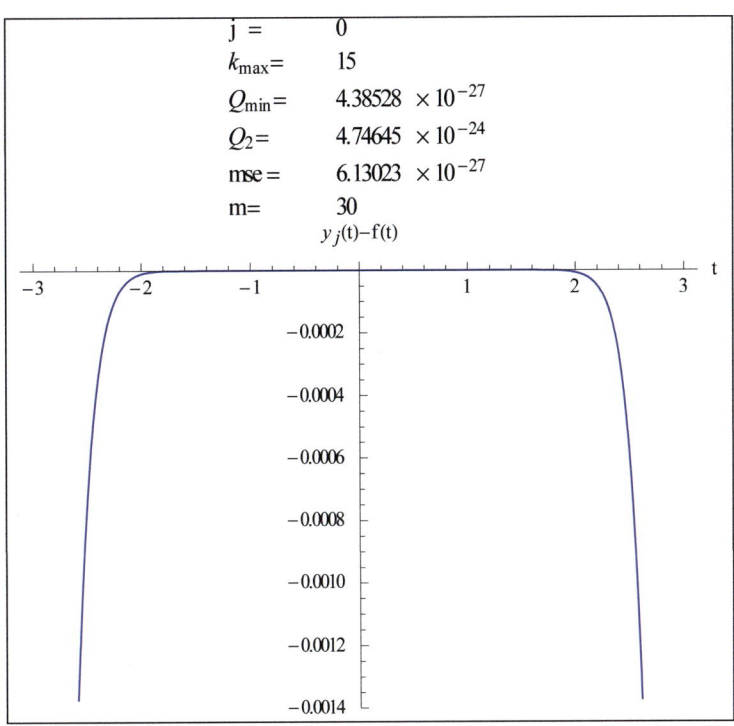

Figure 19. Graph of $y_j - y$ for $j = 0$ und $k_{max} = 15$ on *[-3, 3]*

To what extent one can extrapolate the approximation depends on the number of coefficients $c_k$, that means from $k_{max}$ (and generally from $k_{min}$, too) and from the from the width of the approximation interval I. Here $k_{max}$ can not be chosen arbitrarily large, since only the coefficients $c_k$ can be determined well in which $|\phi_{j,k}|$ is still sufficiently large or by using wavelets with compact support at all nonzero.

# A Comparison of the Shannon Wavelet with the Daubechies Wavelet of Order 8

**Example 3:**
We solve approximately the problem of example 1 on $I = [-1, 1]$ and we minimize $Q$ and use the collocation points $t_i = i \cdot h$ (with $i = 1, 2, ..., m$, $m = c \cdot k_{max}$), with $h = 2/(c \cdot k_{max})$ and $k_{min} = -k_{max}$, $k_{max} = 15, 20, 25$, $c = 1, 2, 3$ and $j = 0, 1, 2$.

We use the Shannon wavelet and for a comparison the Daubechies wavelet of order 8.

If we use generally a Daubechies wavelet of order $g$ with the approximation interval $I = [t_0, t_{end}]$ we can chose $k_{min} = 2^j t_0 - (2g-1)+1$ and $k_{max} = 2^j t_{end} - 1$, because of the compact support of the Daubechies wavelet (otherwise $\phi_{j,k} = 0$ on $I$).

In example 3 we have $g = 8$. Two tables follow for comparing the results:

Daubechies wavelet:

| $j$ | $m$ | $k_{min}$ | $k_{max}$ | $Q_{min}$ | $Q_2$ | $mse$ |
|---|---|---|---|---|---|---|
| 0 | 15. | −15 | 0. | $4.35628 \times 10^{-28}$ | $0.00466786$ | $8.43333 \times 10^{-7}$ |
| 0 | 30. | −15 | 0. | $1.7124 \times 10^{-8}$ | $1.0587 \times 10^{-6}$ | $2.27548 \times 10^{-11}$ |
| 1 | 17. | −16 | 1. | $1.00872 \times 10^{-29}$ | $0.29901$ | $0.0000127997$ |
| 1 | 34. | −16 | 1. | $1.46341 \times 10^{-8}$ | $8.28391 \times 10^{-7}$ | $5.72889 \times 10^{-11}$ |
| 2 | 21. | −18 | 3. | $6.71512 \times 10^{-30}$ | $246.608$ | $0.141992$ |
| 2 | 42. | −18 | 3. | $3.57194 \times 10^{-9}$ | $0.0340748$ | $1.89731 \times 10^{-8}$ |

Shannon:

| $j$ | $m$ | $k_{min}$ | $k_{max}$ | $Q_{min}$ | $Q_2$ | $mse$ |
|---|---|---|---|---|---|---|
| 0 | 15 | −15 | 15 | $8.23083 \times 10^{-12}$ | $8.38252 \times 10^{-11}$ | $3.9383 \times 10^{-14}$ |
| 0 | 30 | −15 | 15 | $6.25902 \times 10^{-27}$ | $1.41041 \times 10^{-24}$ | $2.56371 \times 10^{-27}$ |
| 0 | 20 | −20 | 20 | $1.23927 \times 10^{-11}$ | $6.93437 \times 10^{-11}$ | $1.85861 \times 10^{-14}$ |
| 0 | 40 | −20 | 20 | $1.45232 \times 10^{-26}$ | $9.48622 \times 10^{-25}$ | $2.32583 \times 10^{-27}$ |
| 0 | 25 | −25 | 25 | $2.2285 \times 10^{-11}$ | $7.84257 \times 10^{-11}$ | $1.78816 \times 10^{-14}$ |
| 0 | 50 | −25 | 25 | $1.20324 \times 10^{-26}$ | $6.9879 \times 10^{-25}$ | $4.64306 \times 10^{-27}$ |
| 1 | 15 | −15 | 15 | $2.99289 \times 10^{-18}$ | $6.02327 \times 10^{-12}$ | $3.72281 \times 10^{-15}$ |
| 1 | 30 | −15 | 15 | $9.91007 \times 10^{-30}$ | $3.01769 \times 10^{-25}$ | $3.27535 \times 10^{-29}$ |
| 1 | 20 | −20 | 20 | $4.97124 \times 10^{-9}$ | $8.27043 \times 10^{-8}$ | $1.56287 \times 10^{-11}$ |
| 1 | 40 | −20 | 20 | $1.92217 \times 10^{-27}$ | $2.78109 \times 10^{-24}$ | $3.97155 \times 10^{-28}$ |
| 1 | 25 | −25 | 25 | $8.75935 \times 10^{-10}$ | $1.37052 \times 10^{-8}$ | $6.55506 \times 10^{-12}$ |
| 1 | 50 | −25 | 25 | $5.66866 \times 10^{-27}$ | $3.01211 \times 10^{-23}$ | $2.03996 \times 10^{-27}$ |
| 2 | 15 | −15 | 15 | $2.17737 \times 10^{-29}$ | $0.000289405$ | $1.76241 \times 10^{-7}$ |
| 2 | 30 | −15 | 15 | $1.69325 \times 10^{-28}$ | $6.41826 \times 10^{-18}$ | $1.0083 \times 10^{-21}$ |
| 2 | 20 | −20 | 20 | $9.6976 \times 10^{-8}$ | $0.0000230687$ | $5.39382 \times 10^{-9}$ |
| 2 | 40 | −20 | 20 | $7.86162 \times 10^{-28}$ | $1.77107 \times 10^{-20}$ | $1.3874 \times 10^{-24}$ |
| 2 | 25 | −25 | 25 | $4.2941 \times 10^{-8}$ | $2.35033 \times 10^{-6}$ | $3.54143 \times 10^{-10}$ |
| 2 | 50 | −25 | 25 | $3.08239 \times 10^{-28}$ | $2.38455 \times 10^{-24}$ | $6.62603 \times 10^{-29}$ |

Here are the Graphs of $y_j$ and $y$, $y_j - y$ and of $y_j$ and $y$ on a bigger interval (here [-3, 3]) for an extrapolation outside the approximation interval $I$ an at least the graph of $d$. We start with the Daubechies wavelet:

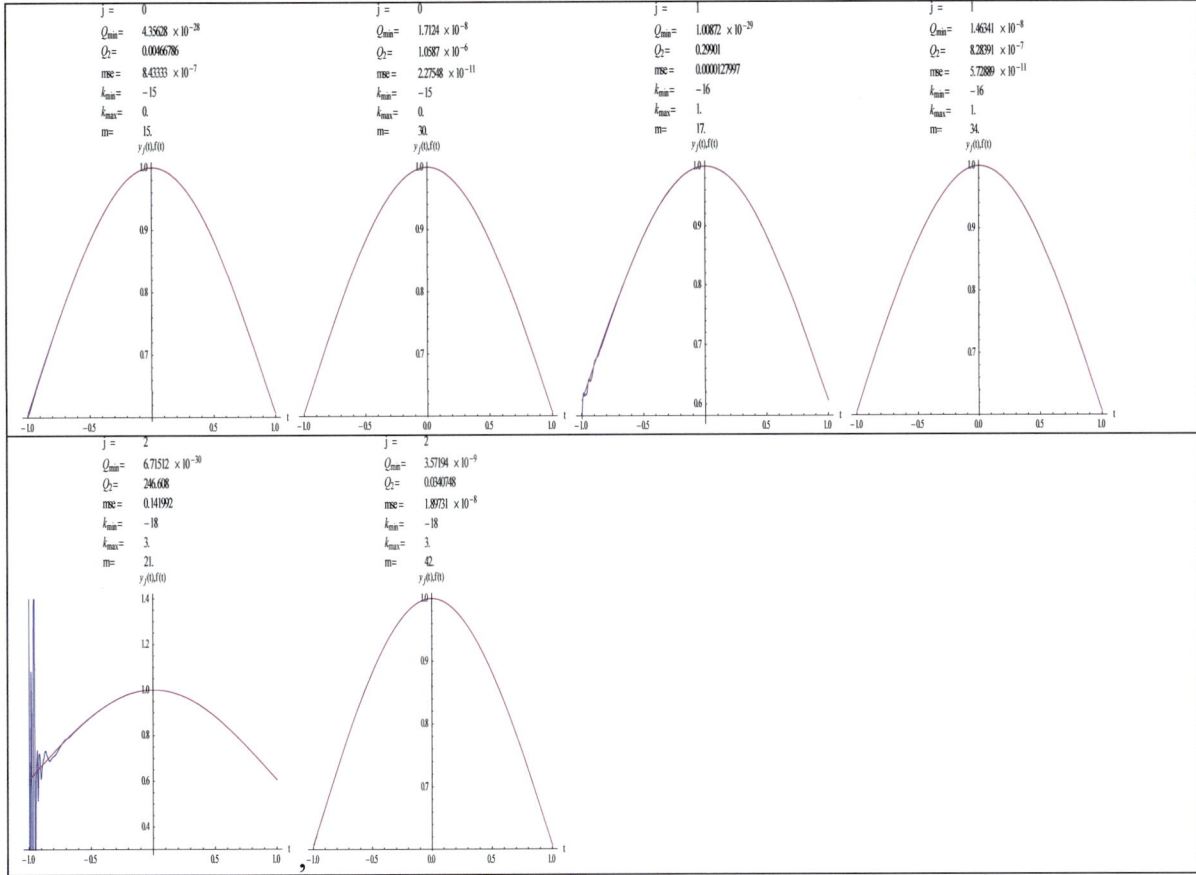

Figure 20. Graphs of $y_j$ and $y$ with the Daubechies wavelet

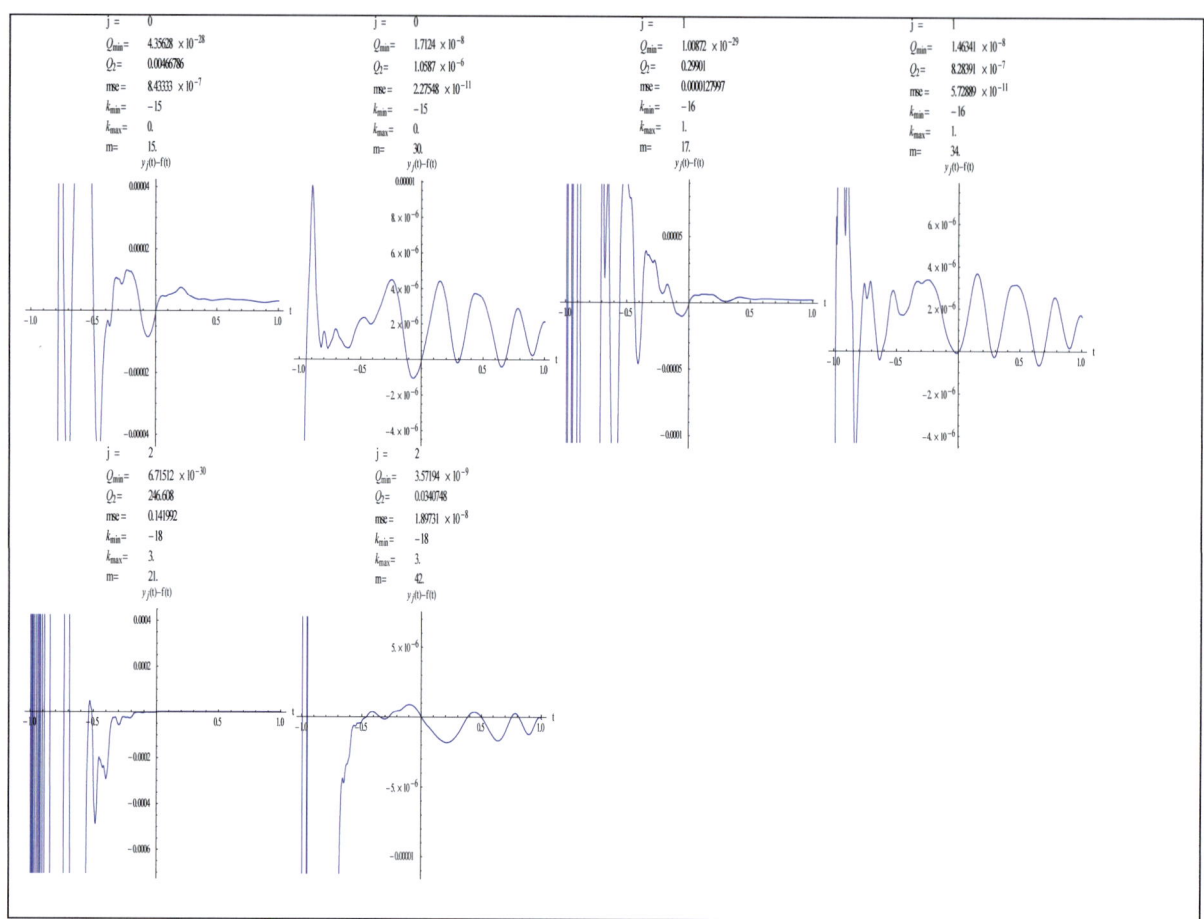

Figure 21. Graphs of $y_j$ - $y$ with the Daubechies wavelet

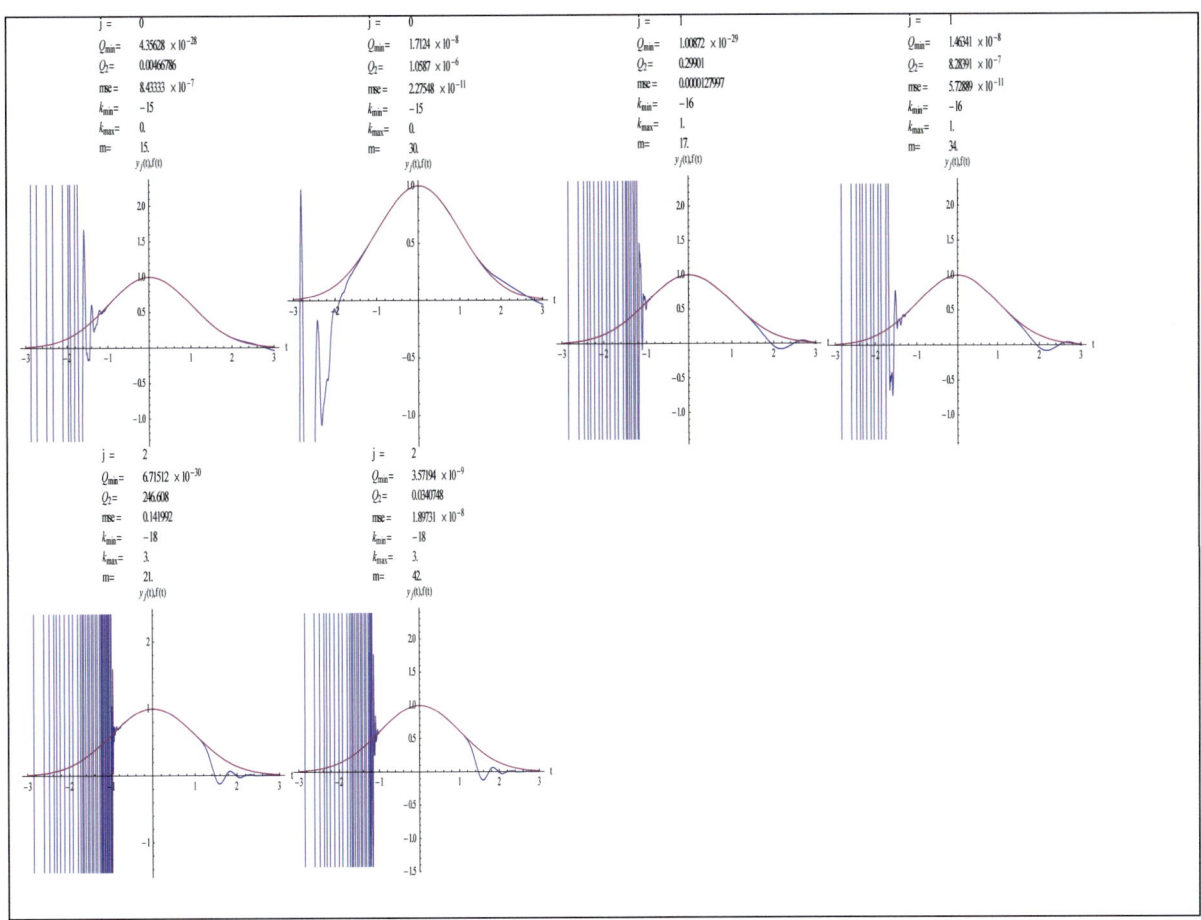

Figure 22. Graphs of $y_j$ and $y$ with the Daubechies wavelet on the interval [-3, 3]

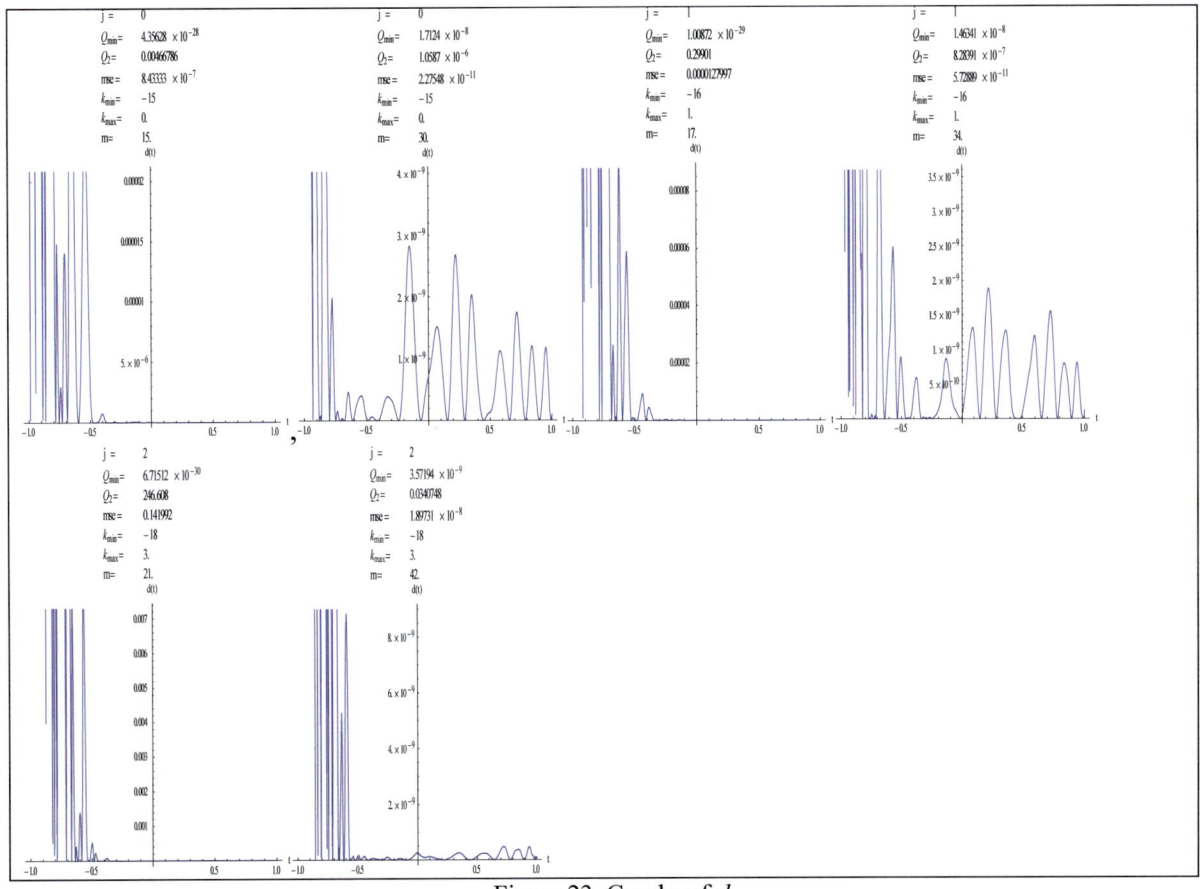

Figure 23. Graphs of $d$

Now the same curves for the Shannon wavelet:

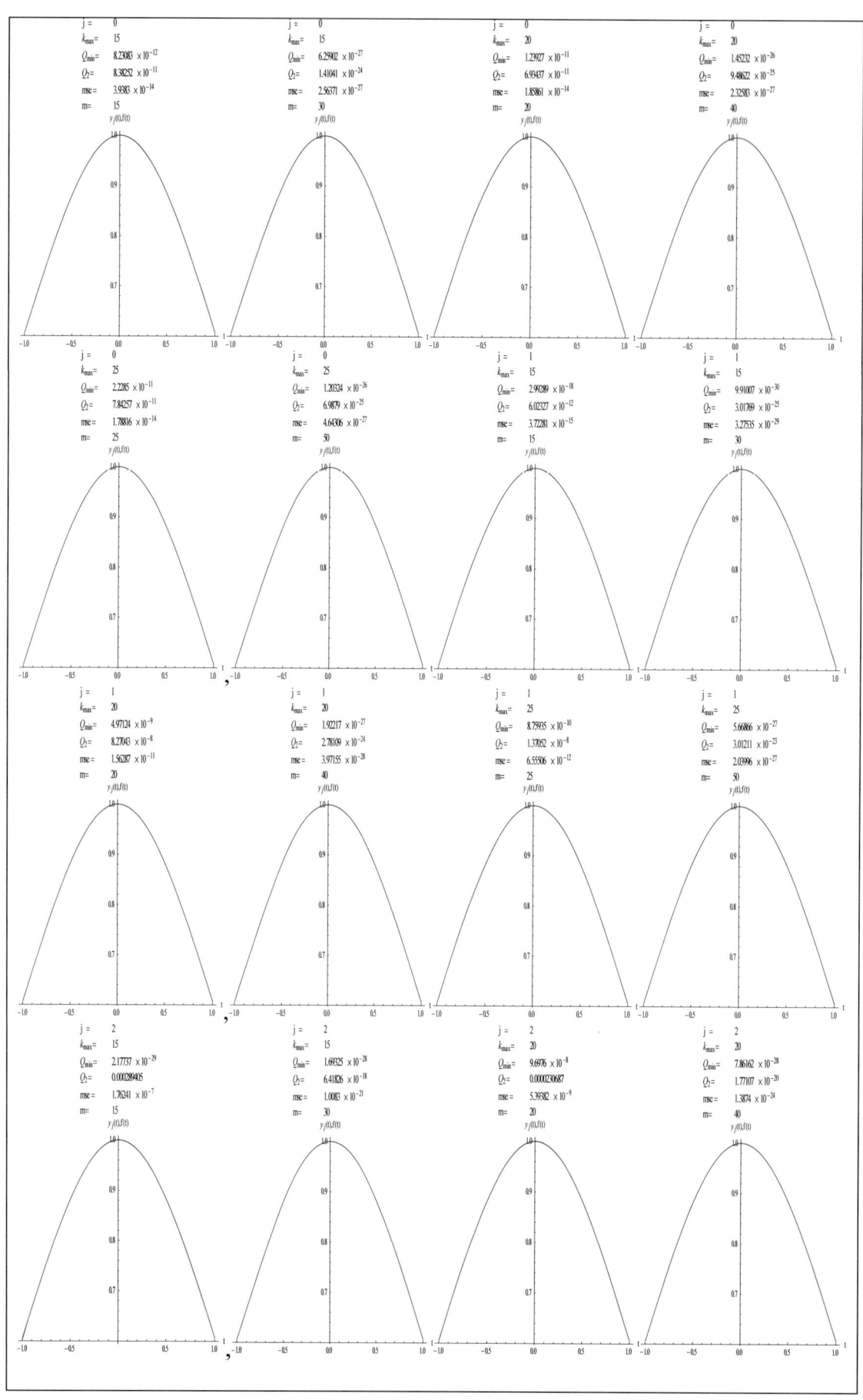

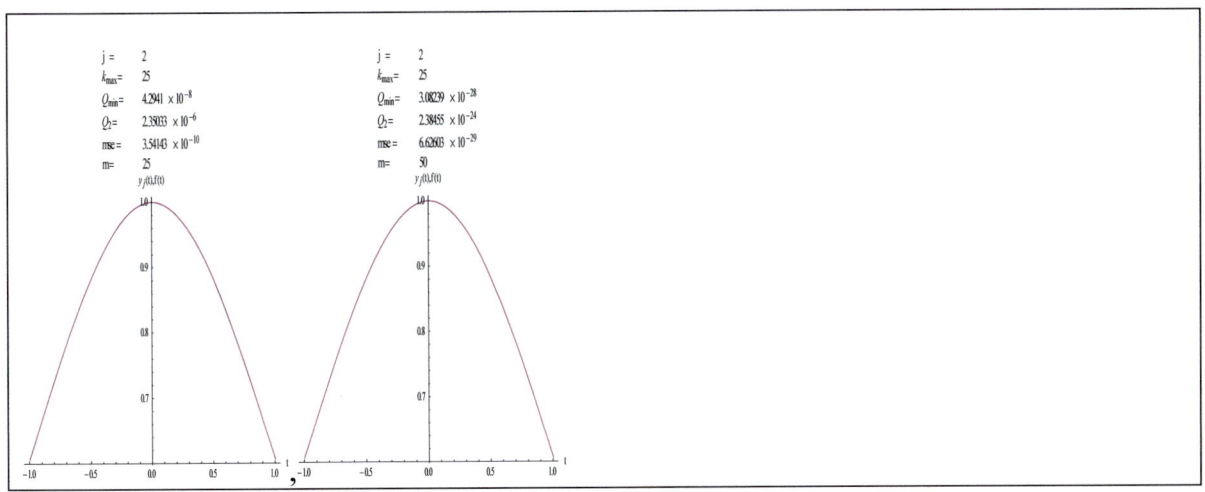

Figure 24. Graphs of $y_j$ and $y$ with the Shannon wavelet

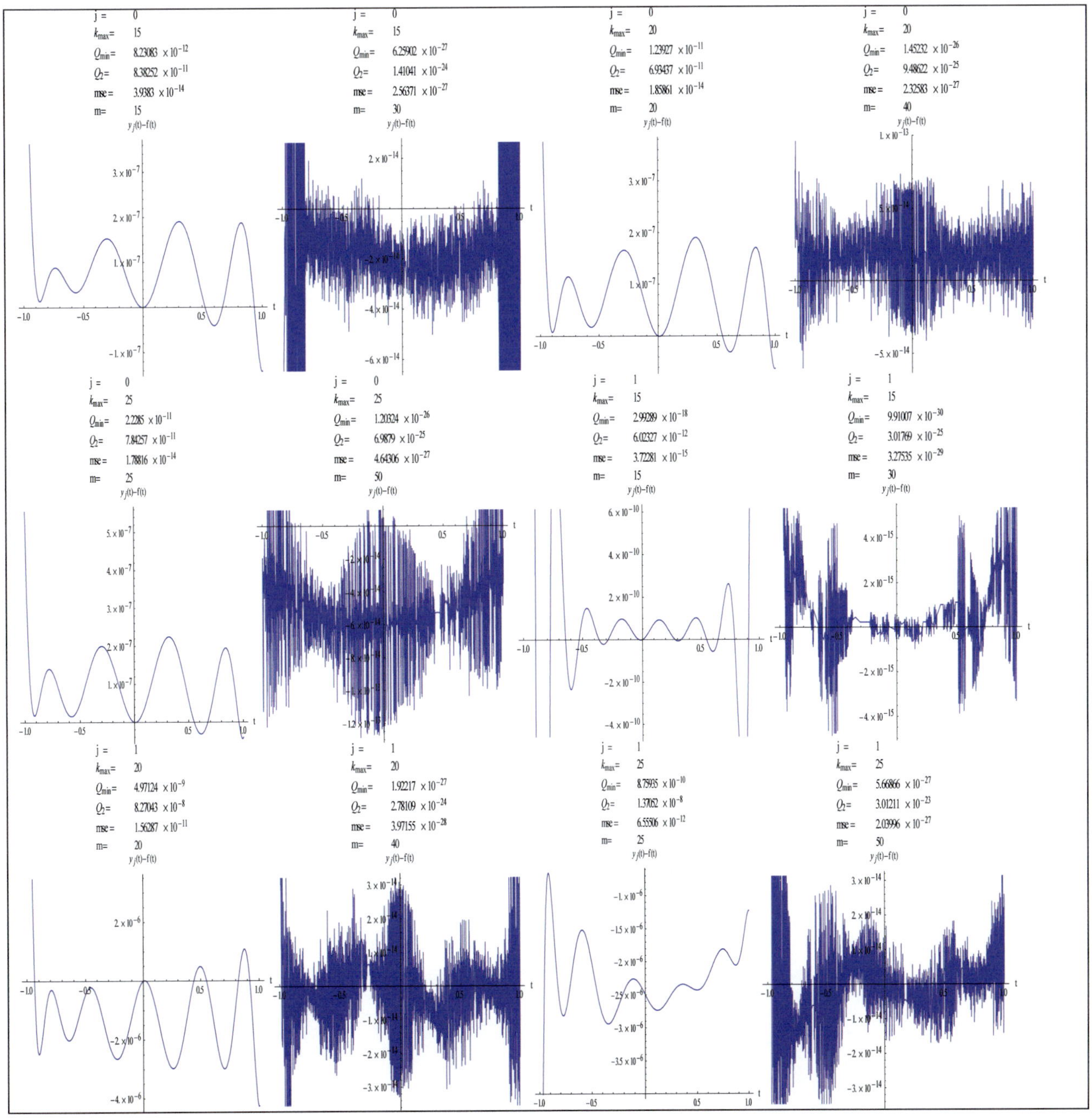

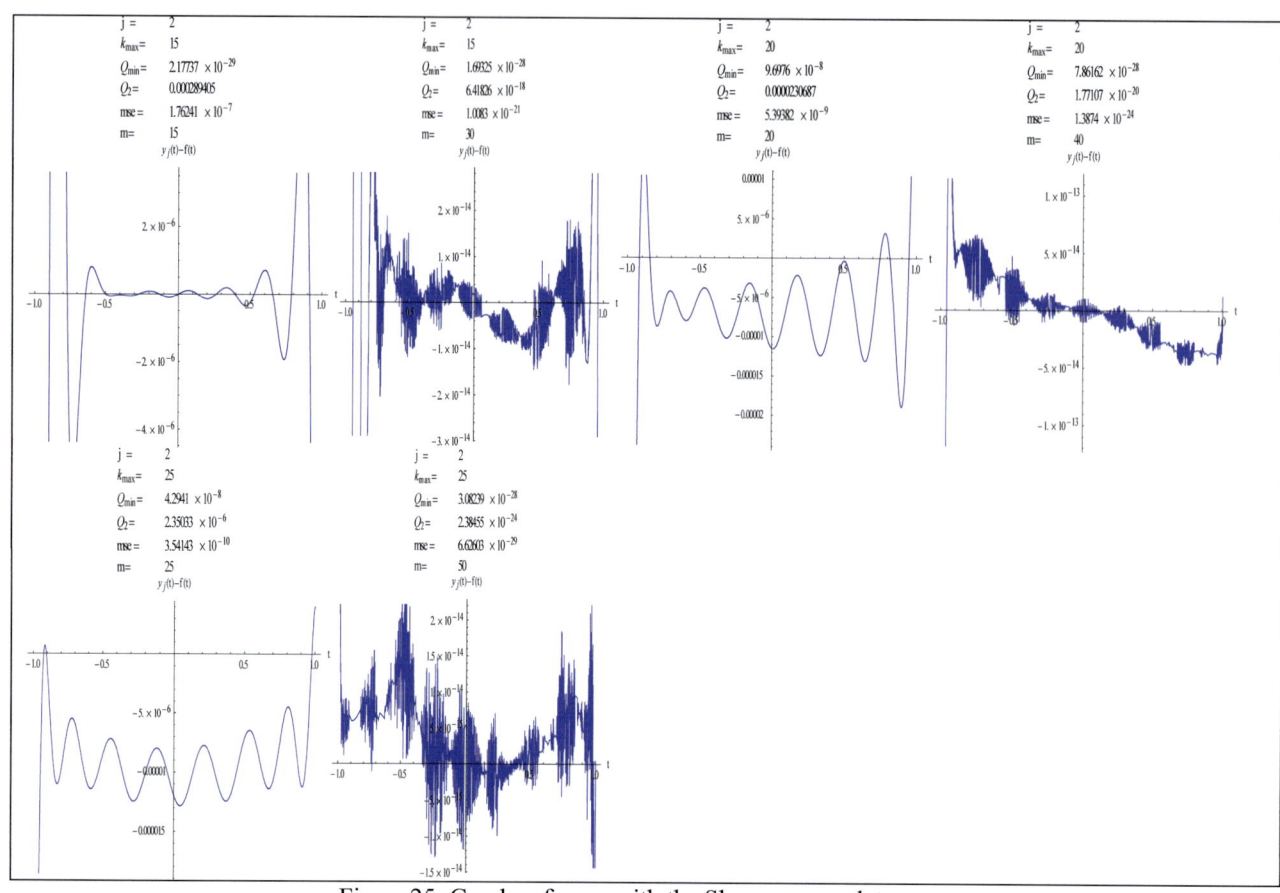

Figure 25. Graphs of $y_j - y$ with the Shannon wavelet

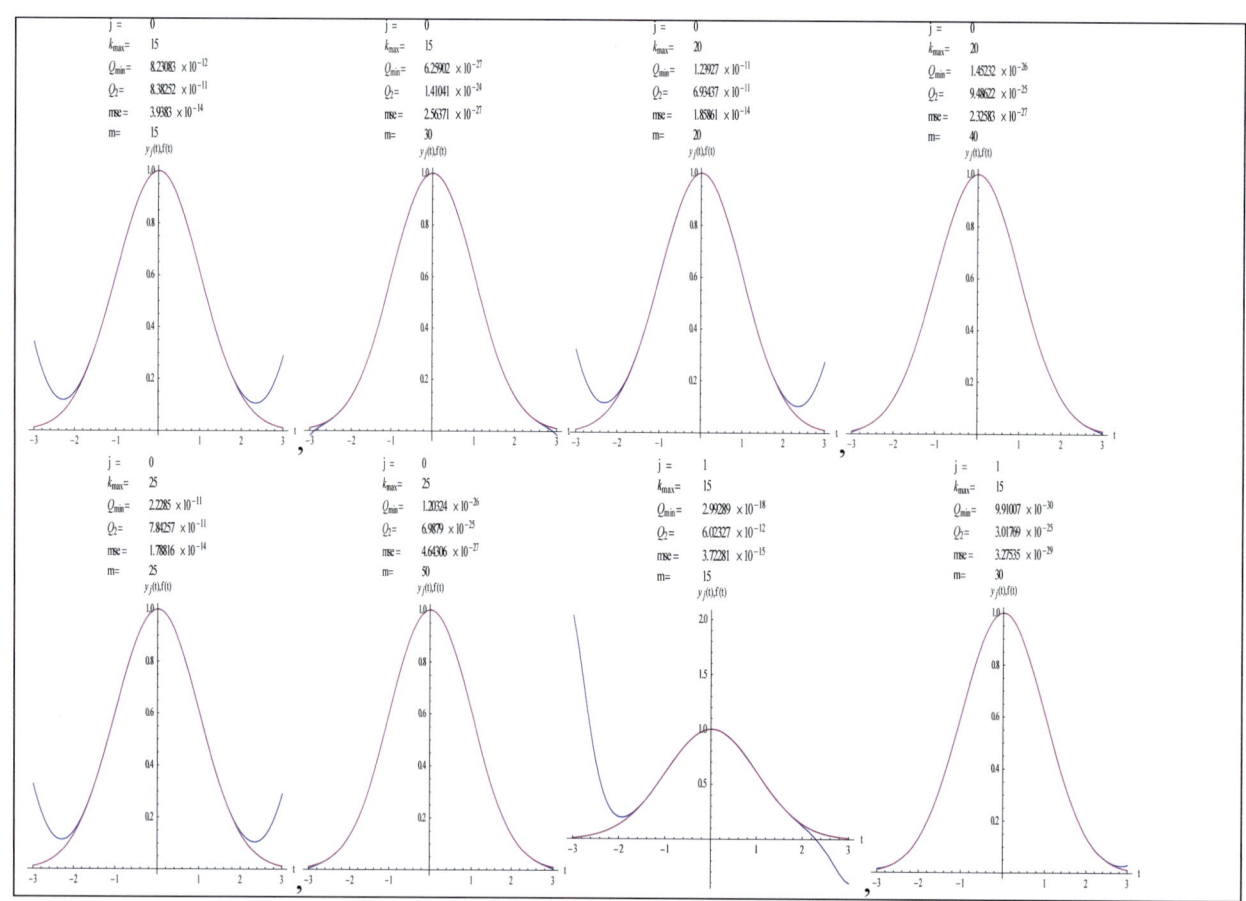

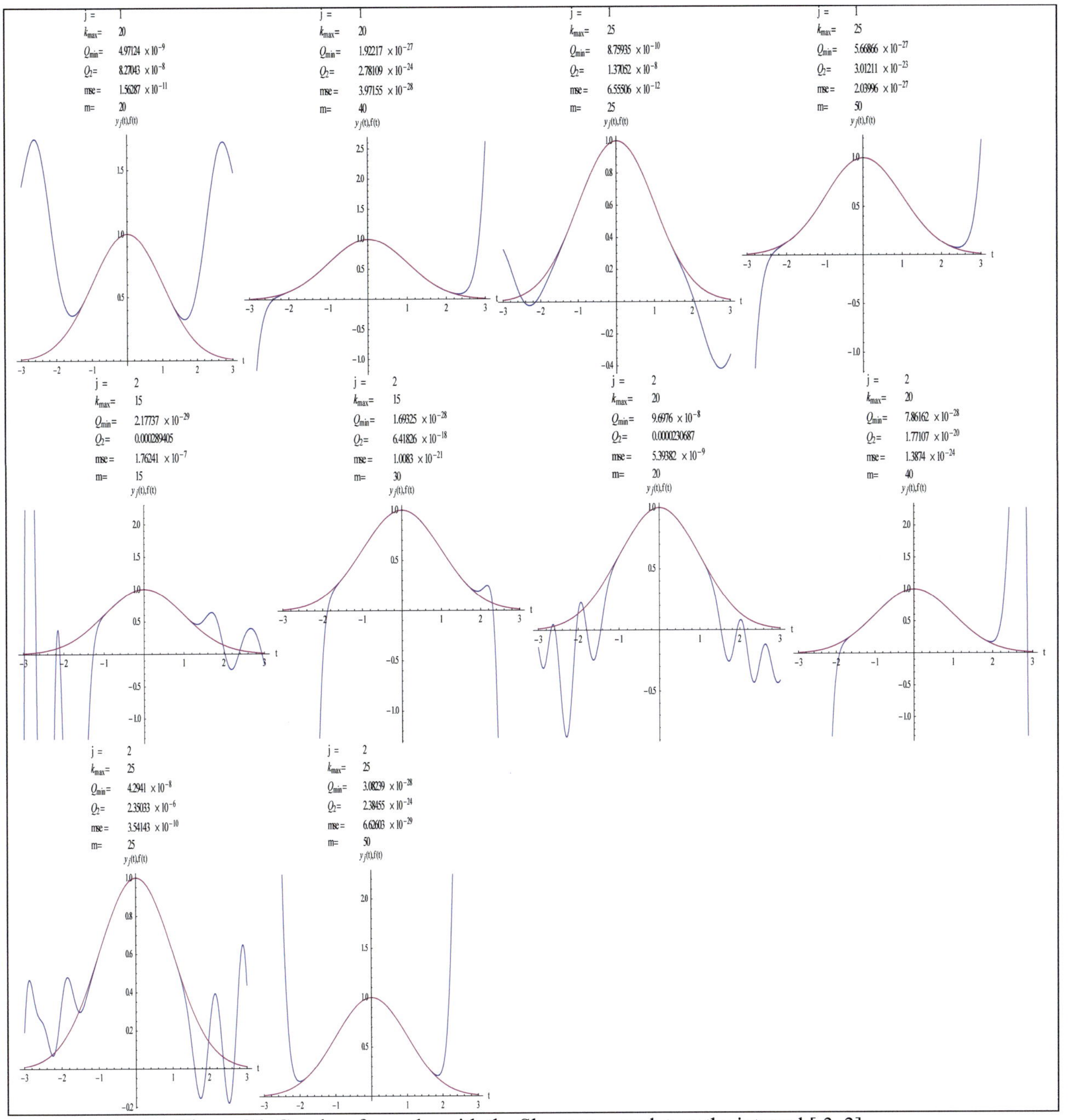

Figure 26. Graphs of $y_j$ and $y$ with the Shannon wavelet on the interval [-3, 3]

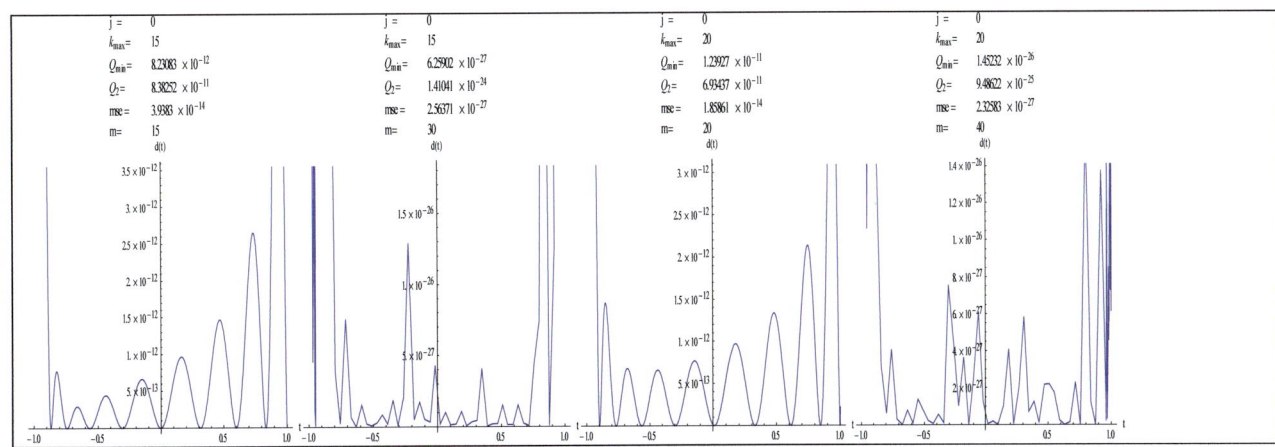

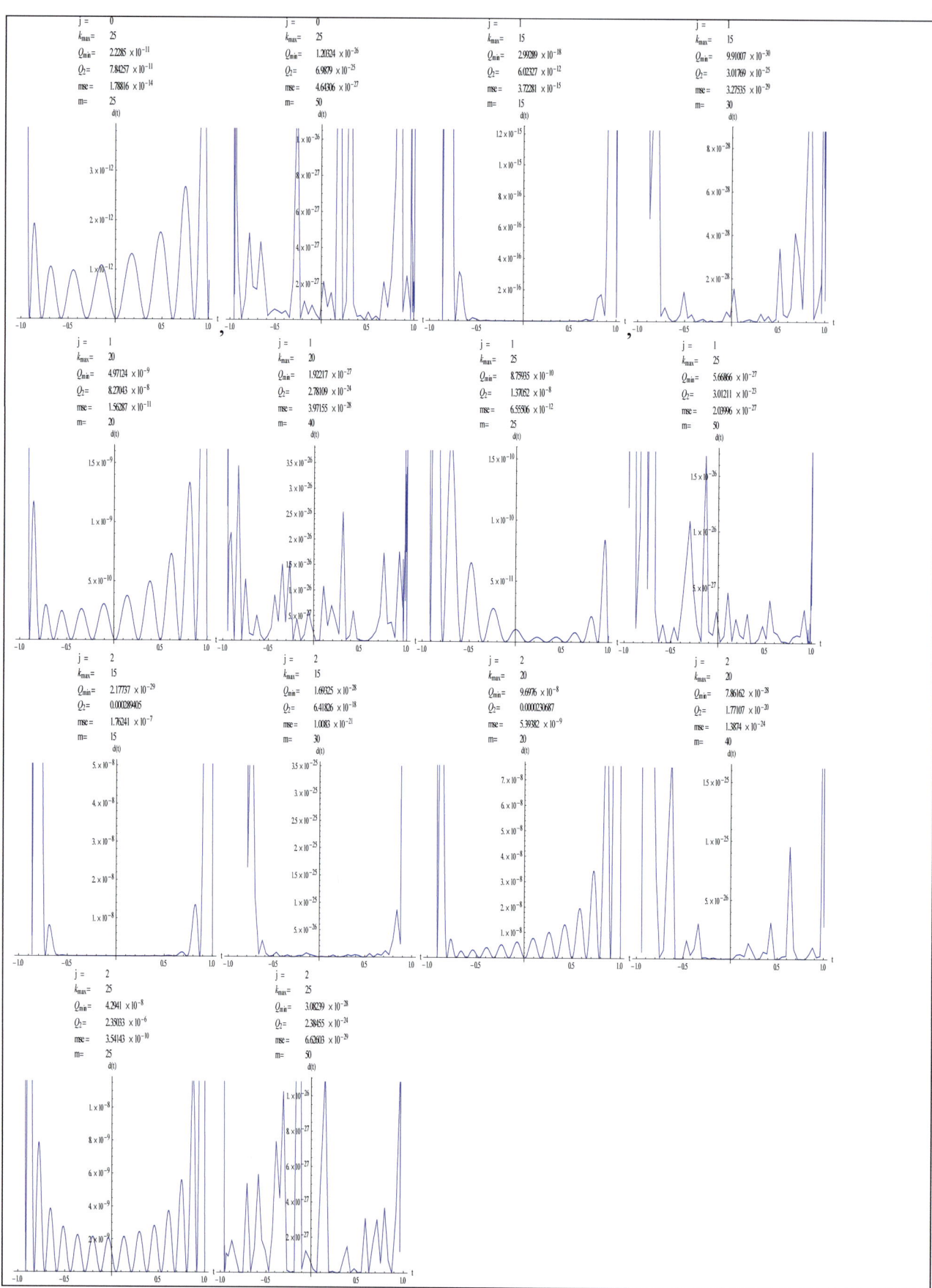

Figure 27. Graphs of $d$ with the Shannon wavelet

The best extrapolation with the Daubechies wavelet:

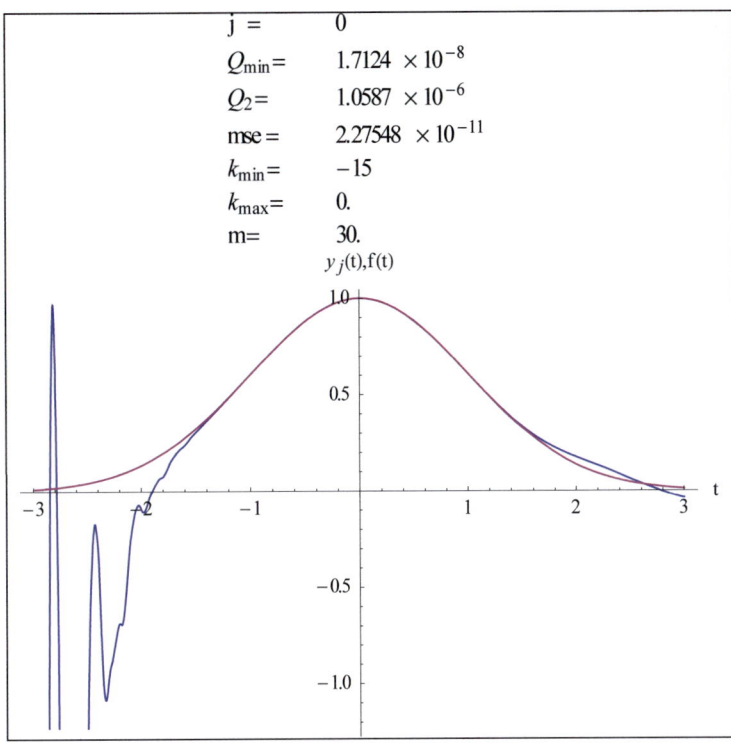

Figure 28. Best extrapolation with the Daubechies wavelet

An extrapolation with the Shannon wavelet:

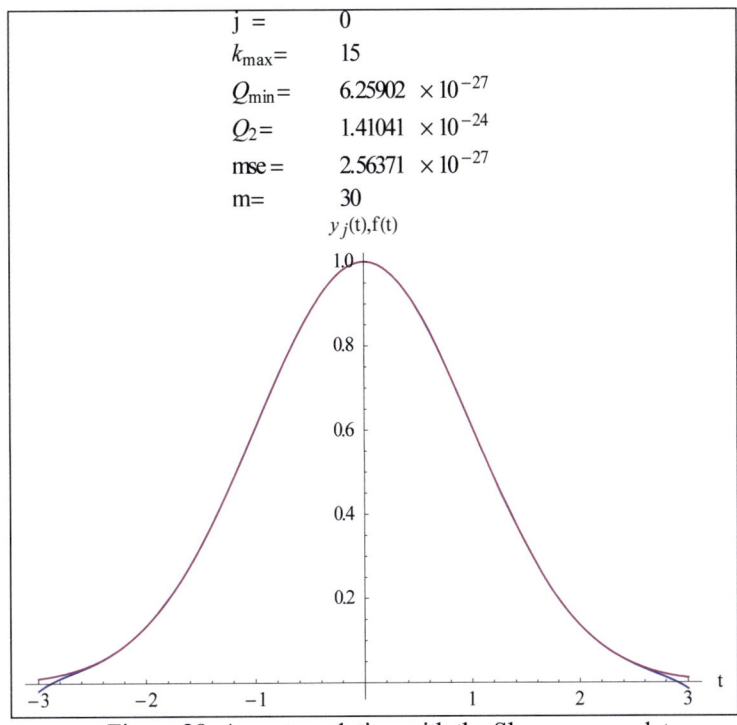

Figure 29. An extrapolation with the Shannon wavelet

With the Shannon wavelet be got much smaller *sse*'s and the extrapolation was much better, too. That is interesting, because the Shannon wavelet has no big order, but here we don't calculate an orthogonal projection on $V_j$. With the Daubechies wavelet we need less coefficients $c_k$. In many simulations we saw that we need a bigger $j$ with the Daubechies wavelets (order 5, 7 and 8) to get an as good approximation as with the Shannon wavelet.

# References

[1] S. Bertoluzza. *Adaptive wavelet collocation method for the solution of Burgers equation.* Transport Theory and Statistical Physics, 25: 3-5, (1996).

[2] T. S. Carlson, J. Dockery, J. Lund. *A Sinc-Collocation Method for Initial Value Problems.* Mathematics and Computation, Vol. 66, No. 217, (1997).

[3] D. L. Donoho. *Interpolating Wavelet Transforms.* Tech. Rept. 408. Department of Statistics, Stanford University, Stanford, (1992).

[4] E. Hairer, G. Wanner. *Vol. 1 : Nonstiff Problems.* Springer, 2. Ed., (1993).

[5] E. Hairer, G. Wanner. *Vol. 2 : Stiff and Differential-Algebraic Problems.* Springer, 2. Ed., (1996).

[6] L. Qian. *On the Regularized Whittaker-Koltel'nikov-Shannon Sampling Theorem.* Proceedings of the Amarican Mathematical Society, Vol. 131, No. 4, (2002).

[7] M. Schuchmann. *Approximation and Collocation with Wavelets. Approximations and Numerical Solving of ODEs, PDEs and IEs.* Osnabrück, DAV, *(2012).*

[8] G. Strang. *Wavelets and Dilation Equations: A Brief Introduction.* SIAM Review Vil. 31, No. 4, (1989)

[9] M. Unser, T. Blu. *Comparison of Wavelets from the Point of View of their Approximation Error.* Proc. of SPIE Vol. 3458, Wavelet Applications in Signal and Image Processing, (1998)

[10] M. Unser (1996). *Vanishing moments and the approximation power of wavelet expansions.* Proceedings of the 1996 IEEE International Conference on Image Processing, (1996).

[11] R. Vuduc. *A Wavelet Collocation Method for Solving PDEs.* J. Comp. Phys., (2000).

# Journal of Approximation Theory and Applied Mathematics

2013 Vol. 2

## Contents

*Solving ODEs and DAEs with a Wavelet Collocation Method with Examples from the Chemical Reaction Kinetics: 3 - 12*

*Solving Integral Equations with a Wavelet Collocation Approach: 13 - 16*

*Approximation of Non $L^2(R)$ Functions on a Compact Interval with a Wavelet Base: 17 - 24*

*Comparing Approximations of a Wavelet Collocation Method of Various Wavelets: 25 - 43*

# Solving ODEs and DAEs with a Wavelet Collocation Method with Examples from the Chemical Reaction Kinetics

M. Schuchmann and M. Rasguljajew from the Darmstadt University of Applied Sciences

## Abstract

In this paper we apply a Wavelet Collocation Method to solve numerically an ODE and a DAE. This Method can be used in multiple cases, even for boundary value problems, PDEs or IEs. The examples we use belongs to the chemical reaction kinetic and the DAE is a test problem, which could be written as a stiff ODE.

## Introduction

In the wavelet theory a scaling function $\phi$ is used wich belongs to a MSA (multi scale analysis with the following properties:

$$\ldots \subset V_{-1} \subset V_0 \subset V_1 \subset \ldots \subset L^2(R),$$

$\{\phi_{j,k}(t)\}_{k \in Z}$ is an orthonormal basis of $V_j$.

We use the following Approximation function

$$y_j(t) := \sum_{k=k_{min}}^{k_{max}} c_k \cdot \phi_{j,k}(t) \quad, \text{with } \phi \in C^1(R).$$

In the following examples it is shown how to recognize a bad approximation without knowing the exact solution.

We use

$$y_j(t) := \sum_{k=k_{min}}^{k_{max}} c_k \phi_{j,k}(t)$$

Since the approximation segment is not alway symmetrical to $t = 0$.

In the following chapters we always determine $c$ by minimizing the function

$$Q(c) = \sum_{i=1}^{m} (y_j'(t_i) - f(y_j(t_i), t_i))^2 + (y_j(t_0) - y_0)^2$$

or

(1) $$Q(c) = \sum_{i=1}^{m} \|y_j'(t_i) - f(y_j(t_i), t_i)\|_2^2 + \|y_j(t_0) - y_0\|_2^2 .$$

in case of a system. Analogous we can apply this method to DEAs, which shows the second example.

If f is a system, we can use

$$y_j(t) = \left( \sum_{k=k_{min}}^{k_{max}} c_{k,1} \phi_{j,k}(t), \sum_{k=k_{min}}^{k_{max}} c_{k,2} \phi_{j,k}(t), \ldots, \sum_{k=k_{min}}^{k_{max}} c_{k,n_f} \phi_{j,k}(t) \right)^T,$$

if $y$ consists of $n_f$ components. For the i-th component of $y$ we use $y_i$, but also $y^{(i)}$ if we want to avoid a confusion with $y_j$.

The 'collocation' points $t_i$ are defined by $t_i = t_0 + i \cdot h$ with

(2) $$h = \frac{t_{end} - t_0}{m} \text{ and } m \geq |k_{max} - k_{min}|.$$

For test purposes in simulations we chose different $m$.

Therefore we can compare

$$Q_{min} = \sum_{i=1}^{m} \left\| y_j'(t_i) - f(y_j(t_i), t_i) \right\|_2^2 + \left\| y_j(t_0) - y_0 \right\|_2^2$$

with

$$Q_a = \sum_{i=1}^{m_a} \left\| y_j'(\tau_i) - f(y_j(\tau_i), \tau_i) \right\|_2^2 + \left\| y_j(t_0) - y_0 \right\|_2^2 \text{ with } \tau_i = t_0 + i \cdot h/a,$$

$m_a = a \cdot m$ provided $a$ is an integer (see examples in the following chapters and [12]). If $Q_a \gg Q_{min}$ than $m$ should be increased.

Generally a good starting value for $m$ is $m = |k_{max} - k_{min}|$ provided nothing is known about the solution. This has been shown in numerous simulations. For functions with extreme slopes or curvatures a bigger m is appropriate.

## Example 1: System in Chemical Reaction Kinetics

in chemical reaction kinetics the Shell problem is given by the following differential equation system:

$$y_1' = -p_1 y_1 y_2 + p_2 y_3 + p_7 y_3 y_7 + p_9 y_6 y_7 - p_8 y_1 y_8 - p_{10} y_1 y_9$$
$$y_2' = -p_1 y_1 y_2 + p_2 y_3 - p_5 y_2 y_5 + p_6 y_6$$
$$y_3' = p_1 y_1 y_2 - p_2 y_3 - p_3 y_3 + p_4 y_4 y_5 - p_7 y_3 y_7 + p_8 y_1 y_8 + p_{11} y_6 y_8 - p_{12} y_3 y_9$$
$$y_4' = p_3 y_3 - p_4 y_4 y_5$$
$$y_5' = p_3 y_3 - p_5 y_2 y_5 - p_4 y_4 y_5 + p_6 y_6$$
$$y_6' = p_5 y_2 y_5 - p_6 y_6 - p_9 y_6 y_7 - p_{11} y_6 y_8 + p_{10} y_1 y_9 + p_{12} y_3 y_9$$
$$y_7' = -p_7 y_3 y_7 - p_9 y_6 y_7 + p_8 y_1 y_8 + p_{12} y_1 y_9$$
$$y_8' = p_7 y_3 y_7 - p_8 y_1 y_8 - p_{11} y_6 y_8 + p_{12} y_3 y_9$$
$$y_9' = p_9 y_6 y_7 + p_{11} y_6 y_8 - p_{10} y_1 y_9 - p_{12} y_3 y_9$$

The following results refer to the start vector:

$$y(0) = (0, 3, 0, 0, 0.01, 0, 1, 0, 0)^T.$$

The Shell-Problem originates from the Shell-Laboratories in Amsterdam (see [19]). The following parameter vector is used:

$$p = (0.299, 0.218, 49.5, 0.000363, 0.962, 47.8, 1000, 900, 700, 1260, 7000, 14000)^T$$

Following setup was chosen: $j = 1$, $k_{min} = -5$, $k_{max} = 20$ approximation interval $I = [0, 5]$. Therefore we have $(20+5+1) \cdot 9 = 234$ coefficients! For collocation points we chose

$$t_i = 1/20 \cdot i, \text{ with } i = 1,...,100$$

whereby $m$ was set to 100.

The iteration (Mathematica function FindMinimum, Version 8) was stopped before the norm of the gradient was smaller than the tolerance 'AccuracyGoal' of Mathematica (because the step size was smaller than the tolerance 'PrecisionGoal').

The results were

$$Q_{min} \approx 1.50724 \cdot 10^{-9} \text{ and } Q_2 \approx 1.56389 \cdot 10^{-9}.$$

The largest deviation was at $t_0$. Without the term $\|y_j(t_0) - y_0\|_2^2$ in $Q$ there would be:

$$Q_{min} \approx 4.80455 \cdot 10^{-12} \text{ and } Q_2 \approx 6.14072 \cdot 10^{-11}.$$

Here the biggest deviation is at $y^{(6)}$:

| i | $|y_j^{(i)}(t_0) - y^{(i)}(t_0)|$ |
|---|---|
| 1 | $2.29951 \times 10^{-9}$ |
| 2 | $5.11224 \times 10^{-10}$ |
| 3 | $2.49557 \times 10^{-8}$ |
| 4 | $5.04493 \times 10^{-15}$ |
| 5 | $1.52255 \times 10^{-7}$ |
| 6 | $0.0000387611$ |
| 7 | $1.41979 \times 10^{-9}$ |
| 8 | $1.41324 \times 10^{-8}$ |
| 9 | $2.25085 \times 10^{-9}$ |

Table 1: $|y_j^{(i)}(t_0) - y^{(i)}(t_0)|$

The following graph of $y_j^{(6)} - y^{(6)}$ in a small section at the beginning of the approximation area ($y$ was numerically calculated using the Mathematica function NDSolve). There it can be seen that the deviation is relatively large in the beginning.

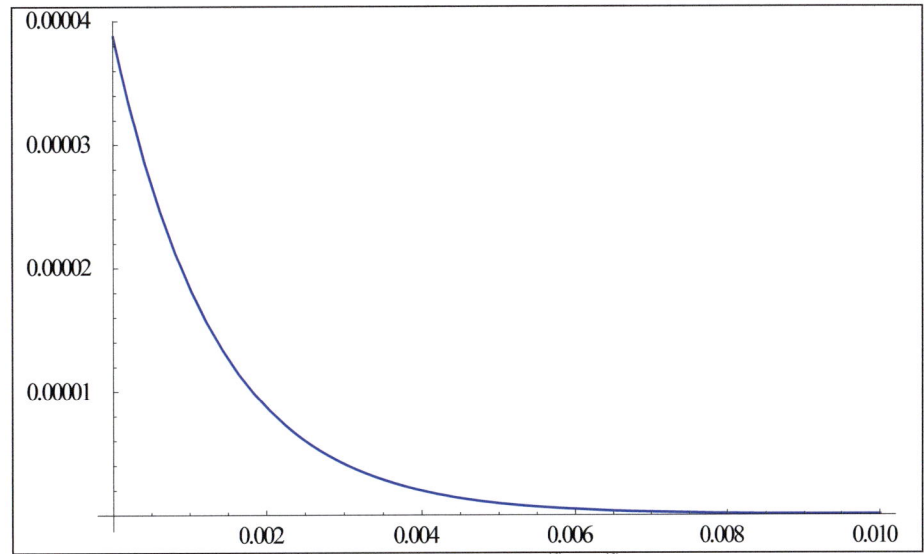

Figure 1. Graph of $y_j^{(6)} - y^{(6)}$

Up next is the graph of $d$ with $d(t) = \|y_j'(t) - f(y_j(t), t)\|_2^2$:

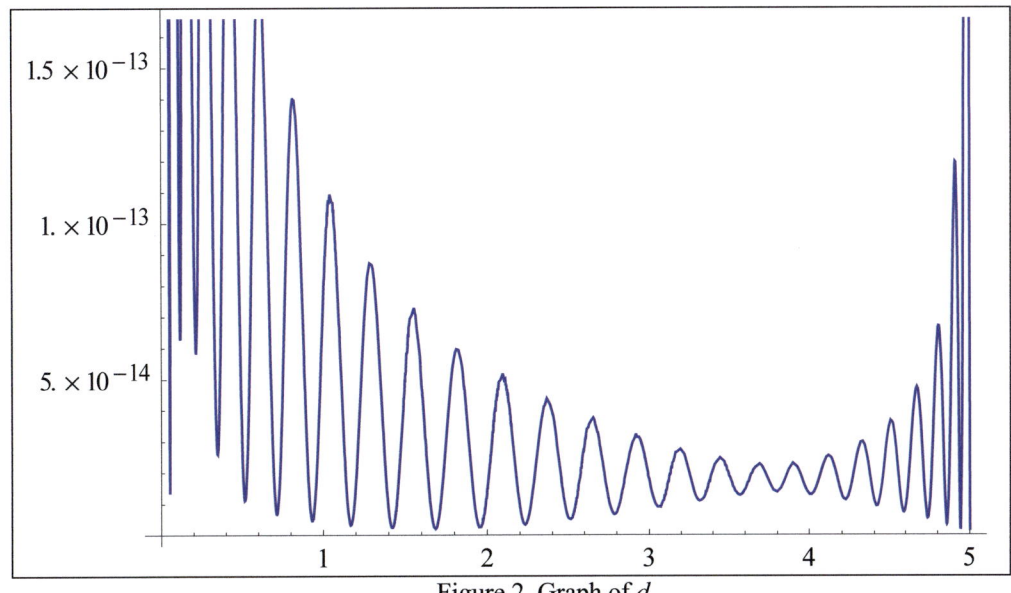

Figure 2. Graph of $d$

Up next the graphs of $y_j^{(i)} - y^{(i)}$:

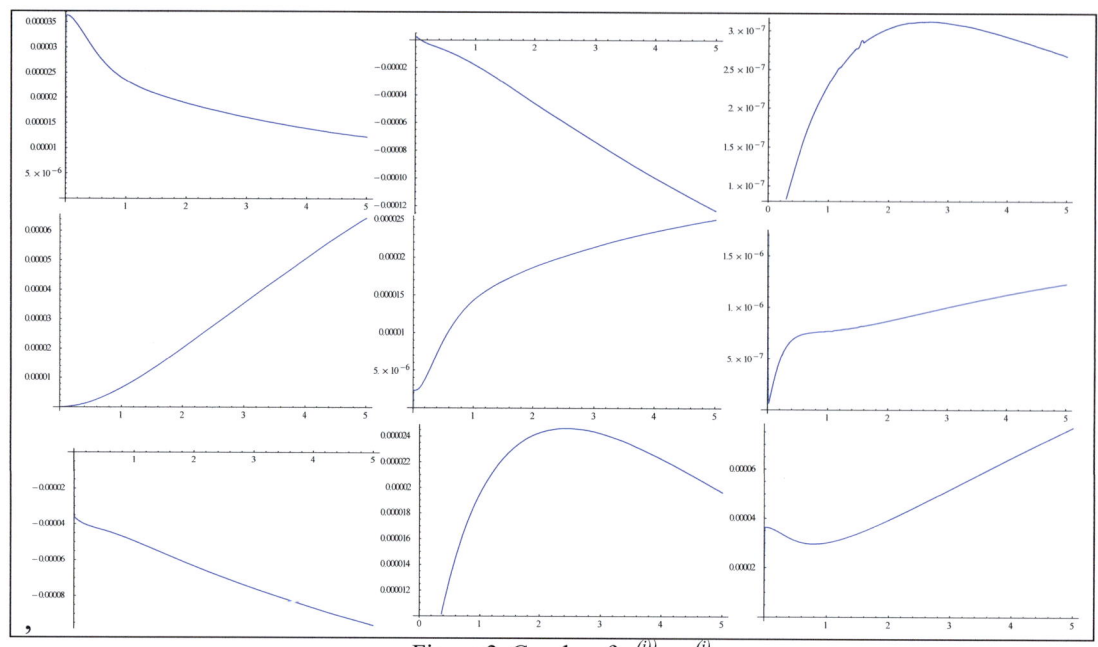

Figure 3. Graphs of $y_j^{(i)} - y^{(i)}$

And finally the graphs of $y_j^{(i)}$ and $y^{(i)}$ (graphically there can no difference be seen):

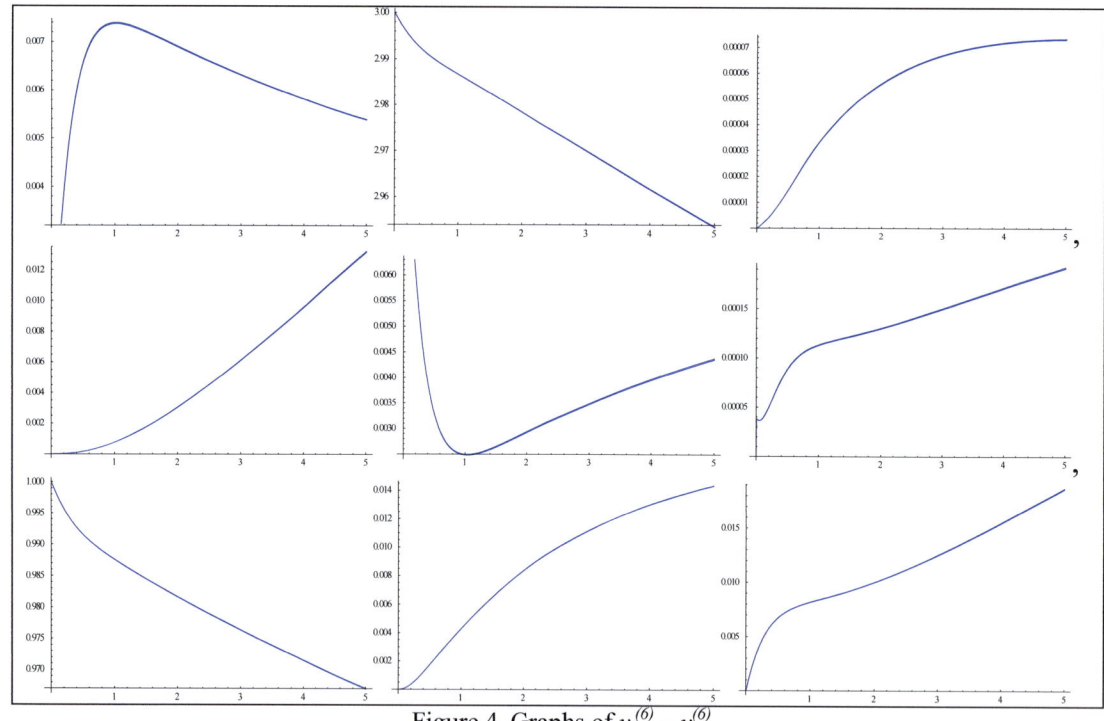

Figure 4. Graphs of $y_j^{(6)} - y^{(6)}$

# Example 2: Differential Algebraic Equation (DAE)

The next example is a differential algebraic equation (DAE) which can be written as a differential equation. This equation is also used later used in parameter identification (see chapter 7.2, example from H.H. Robertson)

From that follows the following system of differential equations

| |
|---|
| $y_1' = -p_1 y_1 + p_3 y_2 y_3$ |
| $y_2' = p_1 y_1 - p_3 y_2 y_3 - p_2 y_2^2$ |
| $1 = y_1 + y_2 + y_3$ |

with $p = (0.04, 3 \cdot 10^7, 10^4)^T$. The starting vector was set to $y(0) = (1, 0, 0)^T$.

The following function is going to be minimized:

$$Q(c) = \sum_{i=1}^{m} \left\| F(y_j'(t_i), y_j(t_i), t_i) \right\|_2^2 + \left\| y_j(t_0) - y_0 \right\|_2^2.$$

At it $F(y'', y', t_i) = (y_1' + p_1 y_1 - p_3 y_2 y_3, y_2' - p_1 y_1 + p_3 y_2 y_3 + p_2 y_2^2, 1 - y_1 - y_2 - y_3)^T$.

Here it is shown that the method of approximation using wavelet bases can generally be applied to differential algebraic equations or boundary value problems.

In this problem $y_2$ has an extreme curvature in the beginning (that can be seen in the following graph at [0, 0.1]).

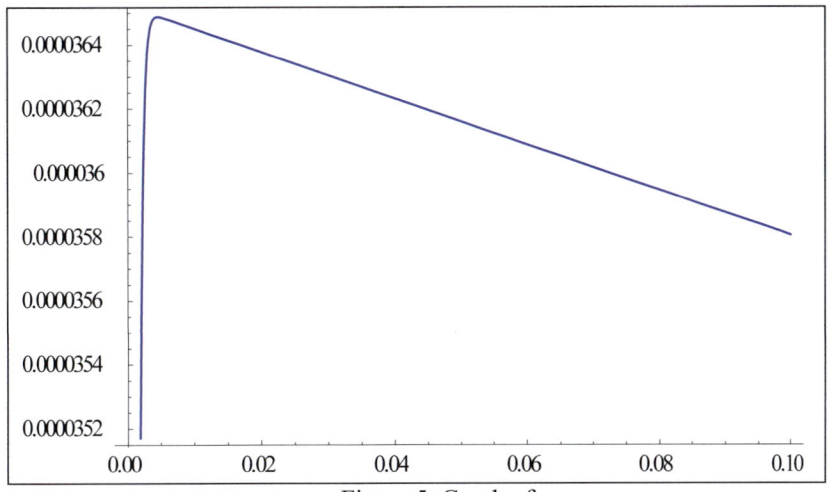

Figure 5. Graph of $y$

Following setup was chosen: $j = 1$, $k_{min} = -5$, $k_{max} = $ approximation interval $I = [0, 5]$.

The iteration (Mathematica funktion FindMinimum, Version 8) was stopped before the norm of the gradient was smaller than the tolerance 'AccuracyGoal' of Mathematica (because the step size was smaller than the tolerance 'PrecisionGoal').

The results were

$$Q_{min} \approx 1.33891 \cdot 10^{-9} \text{ and } Q_2 \approx 1.46643 \cdot 10^{-9}.$$

The largest deviation was at $t_0$. Without the term $\|y_j(t_0) - y_0\|_2^2$ in $Q$ there would be:

$$Q_{min} \approx 5.42071 \cdot 10^{-12} \text{ and } Q_2 \approx 1.32919 \cdot 10^{-10}.$$

Up next is the graph of $d$ with $d(t) = \|F(y_j'(t),(y_j(t),t)\|_2^2$:

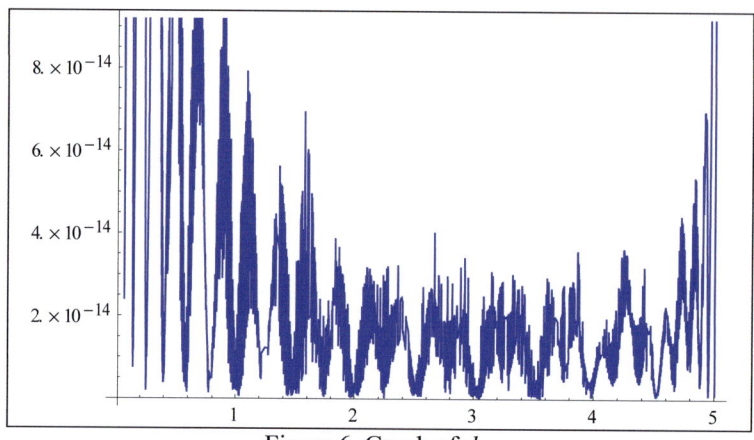

Figure 6. Graph of $d$

Below is the graph of $y_j^{(i)} - y^{(i)}$ ($y$ was numerically calculated using the Mathematica function NDSolve).

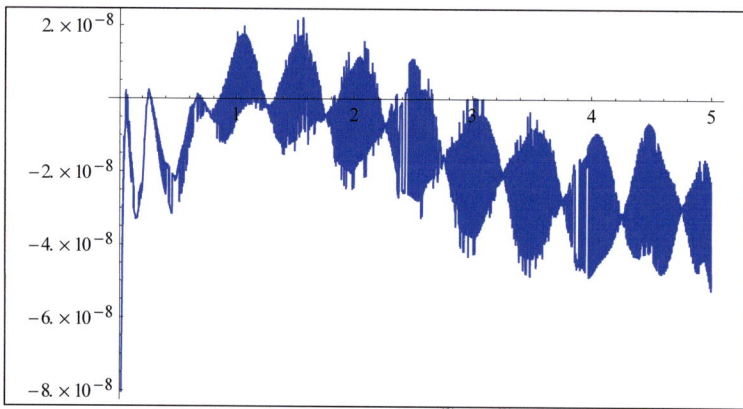

Figure 7. Graph of $y_j^{(1)} - y^{(1)}$

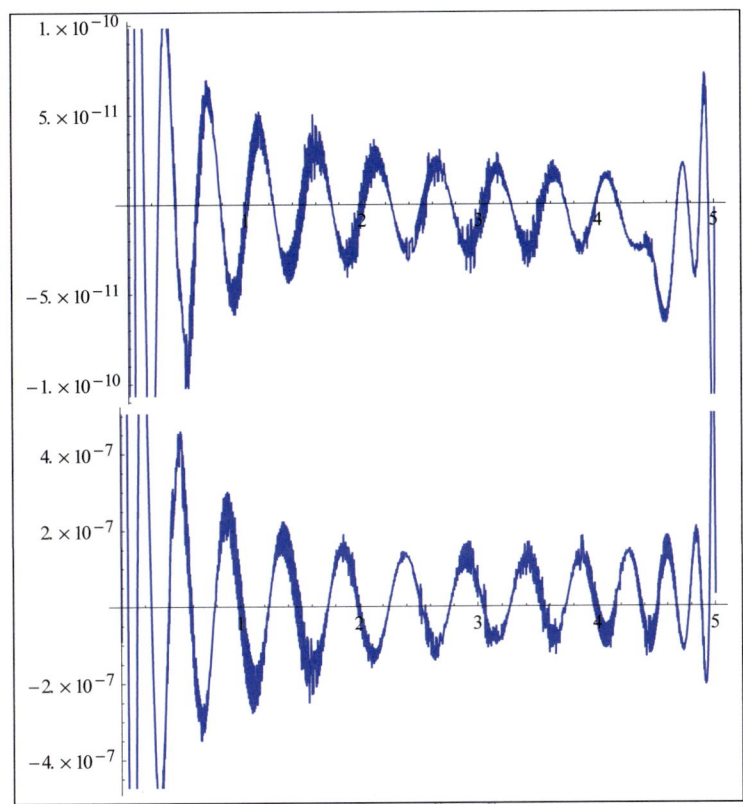

Figure 7. Graphs of $y_j^{(i)} - y^{(i)}$ for $i = 2,3$

And finally the graphs of $y_j^{(i)}$ and $y^{(i)}$ (graphically there can no difference be seen):

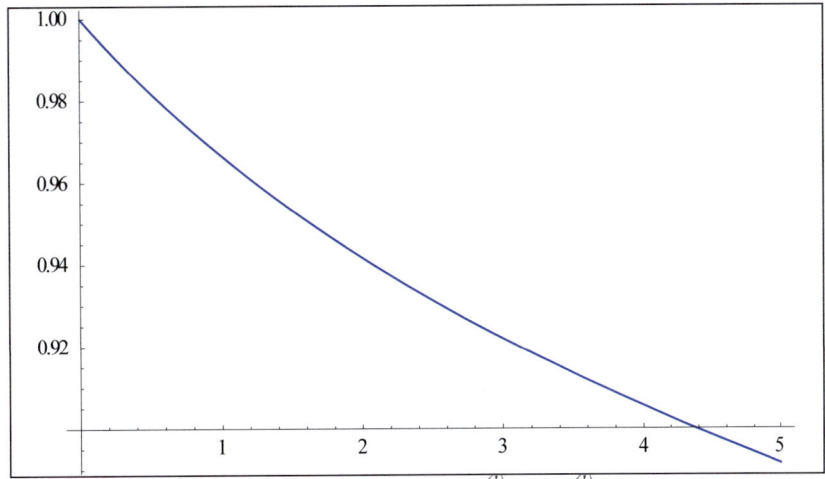

Figure 8. Graphs of $y_j^{(1)}$ and $y^{(1)}$

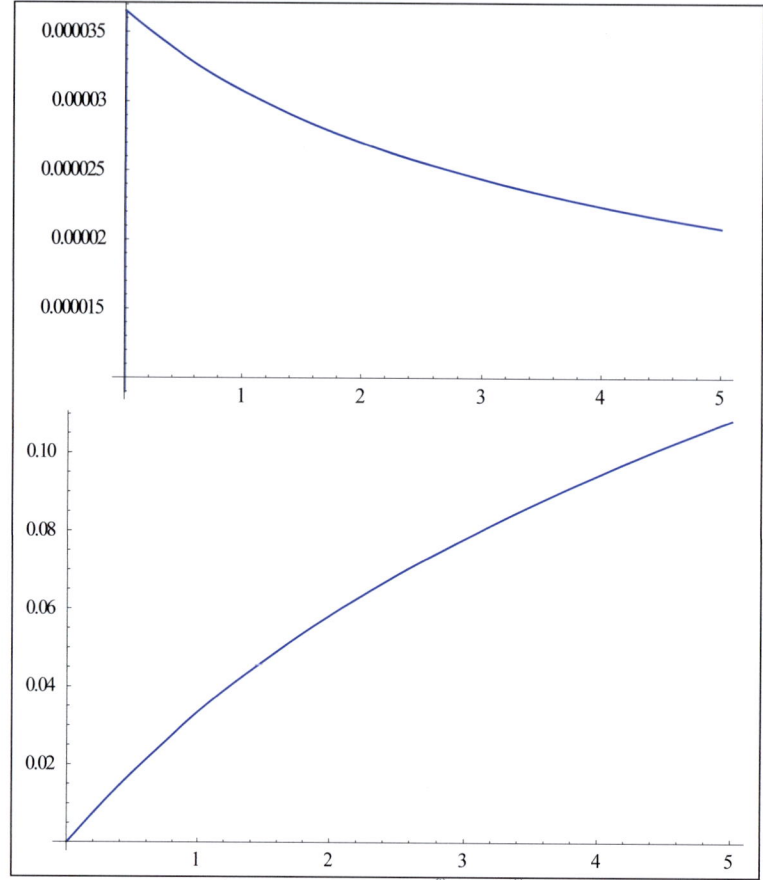

**Figure 9.** Graphs of $y_j^{(i)}$ and $y^{(i)}$ for $i = 2,3$

# References

[1] Abdella, K. (2012). "Numerical Solution of Two-Point Boundary Value Problems Using Sinc Interpolation", *Proceedings of the American Conference on Applied Mathematics (American-Math '12): Applied Mathematics in Electrical and Computer Engineering*

[2] Ascher, U. A. Mattheij, R. M. M. Russell, R. D. (1988). „Numerical Solution of Boundary Value Problems for ODEs", *Prentice Hall (Series in Computational Mathematics)*

[3] Ascher, U. Christiansen, J. Russell, R. (1981). "Collocation Software for Boundary Value ODEs", *ACM Trans. Math. Software*

[4] Bertoluzza S. (2006). "Adaptive Wavelet Collocation Method for the Solution of Burgers Equation," *Transport Theory and Statistical Physics*

[5] Carlson, T. S. Dockery, J. Lund, J. (1997). "A Sinc-Collocation Method for Initial Value Problems", *Mathematics and Computation, Vol. 66, No. 217*

[6] Donoho, D. L.; (1992). "Interpolating Wavelet Transforms," *Tech. Rept. 408. Department of Statistics, Stanford University, Stanford*

[7] Mei, S.-L. Lv, H.-L. Ma, Q. (2008). „Construction of Interval Wavelet Based on Restricted Variational Principle and Its Application for Solving Differential Equations", *Hindawi Publishing Corporation Mathematical Problems in Engineering*

[8] Robertson, H. H. (1975). "Some Properties of Algorithms for Stiff Differential Equations", *J. Inst. Math. Applics.*

[9] Russell, R. D. Christiansen, J. (1979). "A Collocation Solver for Mixed Order Systems of Boundary Value Problems", *Mathematics of Computation*

[10] Schuchmann, M. (2012). "Approximation and Collocation with Wavelets. Approximations and Numerical Solving of ODEs, PDEs and IEs," *Osnabrück: DAV*

[11] Schuchmann, M. (2008). "Parameteridentifikation dynamischer Systeme auf günstigen Pfaden" (German), *DAV*

[12] Schuchmann, M.; Rasguljajew, M. (2013). Error Estimation of an Approximation in a Wavelet Collocation Method. *Journal of Applied Computer Science & Mathematics, No. 14 (7) / 2013, Suceava*

[13] Schuchmann, M.; Rasguljajew, M. (2013). Parameter Identification with a Wavelet Collocation Method in a Partial Differential Equation. *Journal of Approximation Theory and Applied Mathematics (JATAM) Vol. 1*

[14] Schuchmann, M.; Rasguljajew, M. (2013). An Approach for a Parameter Estimation with a Wavelet Collocation Method. *Journal of Approximation Theory and Applied Mathematics (JATAM) Vol. 1*

[15] Shi, Z.; Kouri, D.J.; Wei, G.W.; Hoffman, D. K.; (1999). „Generalized Symmetric Interpolating Wavelets", *Computer Physics Communications*

[16] Strang, G.; (1989). "Wavelets and Dilation Equations: A Brief Introduction", *SIAM Review Vol. 31, No. 4*

[17] Unser, M. (1996). "Vanishing Moments and the Approximation Power of Wavelet Expansions", *Proceedings of the 1996 IEEE International Conference on Image Processing*

[18] Unser, M. Blu, T. (1998). "Comparison of Wavelets from the Point of View of their Approximation Error", *Proc. Of SPIE Vol. 3458, Wavelet Applications in Signal and Image Processing*

[19] Vasilyev, O. V.; Bowman, C.; (2000). "Second-Generation Wavelet Collocation Method for the Solution of Partial Differential Equations", *Academic Press*

Journal of Approximation Theory and Applied Mathematics, 2013 Vol. 2

## *Solving Integral Equations with a Wavelet Collocation Approach*

M. Schuchmann and M. Rasguljajew from the Darmstadt University of Applied Sciences

### Abstract

In this article we describe how to approximate the solution of an integral equation using a wavelet basis. The same idea of constructing an approximation function with a wavelet basis we examined by solving ODEs und PDEs. In the example we use the Shannon wavelet.

### INTRODUCTION

In the wavelet theory a scaling function $\phi$ is used, which has properties that are defined in the MSA (multi scale analysis). Through the MSA we know, that we can construct an orthonormal basis of a closed subspace $V_j$, where $V_j$ belongs to a sequence of subspaces with the following property:

$$... \subset V_{-1} \subset V_0 \subset V_1 \subset ... \subset L^2(\mathbb{R}),$$

$\{\phi_{j,k}(t)\}_{k \in \mathbb{Z}}$ is an orthonormal basis of $V_j$ with $\phi_{j,k}(t) = 2^{j/2}\phi(2^j t - k)$.

We use the following approximation function to approximate the solution of integral equations (EQ):

$$y_j(t) := \sum_{k=k_{min}}^{k_{max}} c_k \cdot \phi_{j,k}(t) \quad .$$

$k_{max}$ and $k_{min}$ depend on the approximation interval $[t_0, t_{end}]$.

The idea is to replace the unknown function in an integral equation trough the approximation function $y_j$. Then we have $m := |k_{max} - k_{min}| + 1$ unknown coefficients $c_k$. The same idea is used for solving ODEs with the collocation method. Then we construct a system of equations for the unknown coefficients, where the integral equation with the substituted approximation function should be fulfilled at $m$ discrete (for example equidistant) points in the integration area of the integral equation.

If we have for example an initial value problem

$$y' = f(y,t)$$
$$y(t_0) = y_0$$

we can write it as an integral equation:

$$y(t) = y_0 + \int_{t_0}^{t} f(y(\tau), \tau) d\tau$$

If we want to approximate the solution y on the interval $I = [t_0, t_{end}]$.

We use the collocation points $t_i$, with $t_i = t_0 + i \cdot h$ ($i = 1, 2, .., m$) and

$$h = \frac{t_{end} - t_0}{m}.$$

Then we can solve the following equations for the $c_k$:

$$y(t_i) = y_0 + \int_{t_0}^{t_i} f(y(\tau), \tau) d\tau \quad \text{, with } i = 1, 2, .., m.$$

If $f$ is linear in $y$ we get a linear system of equations, for example $f(y(t), t) = h(t) \cdot y(t)$, we get

$$y(t_i) = y_0 + \int_{t_0}^{t_i} h(\tau) \cdot \sum_{k=k_{min}}^{k_{max}} c_k \cdot \phi_{j,k}(\tau) d\tau \text{, with } i = 1, 2, .., m$$

and so:

$$y(t_i) = y_0 + \sum_{k=k_{min}}^{k_{max}} c_k \cdot \int_{t_0}^{t_i} h(\tau) \cdot \phi_{j,k}(\tau) d\tau \text{, with } i = 1, 2, .., m.$$

Analogues we can solve numerically other types of integral equations. Now we solve approximately an integral equation in the following example.

**Example:**
The integral equation is

(1) $$y(t) = t + \int_0^t y(\tau) \cdot \sin(t - \tau) d\tau$$

with the solution $\quad y(t) = t + \frac{1}{6} t^3$.

Here we set (with Shannon's $\phi$) (with $k_{max} = -k_{min} = 10$ und $j = 1$)

$$y_1(t) = \sum_{k=-10}^{10} c_k \cdot \phi_{1,k}(t)$$

in the integral equation (1) and we solved the equations

$$y_1(x) = x + \int_0^x y_1(t) \cdot \sin(x - t) dt$$

with $x = 0, \frac{1}{5}, \frac{2}{5}, ..., 4$ for the coefficients $c_{-10}, ..., c_{10}$ with numerical evaluation of the integrals.

Here is the graph of $y_1$:

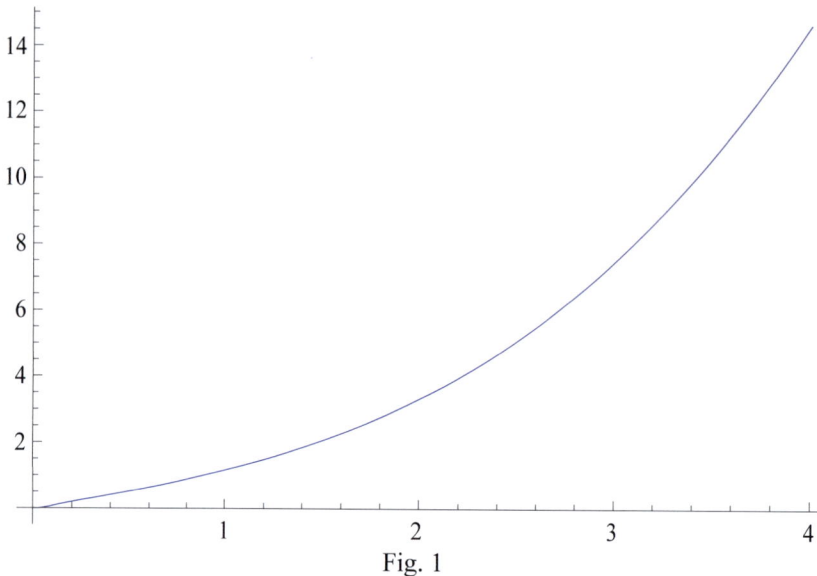

Fig. 1

And the graph of $y_1 - y$:

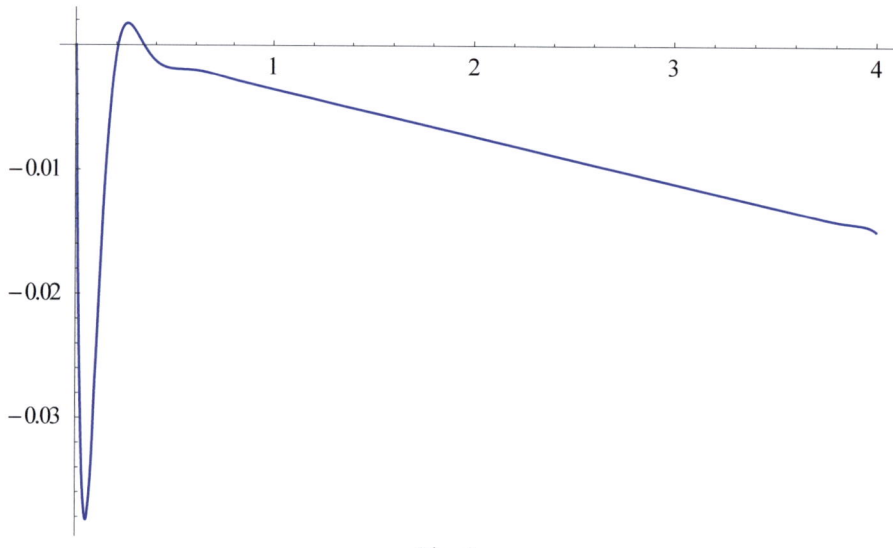

Fig. 2

Now we solve the following minimum problem instead of the equations above and use :

$$\min_{c_k} Q(\vec{c}) = \min_{c_k} \sum_{i=0}^{40} \left(-y_1(i/10) + i/10 + \int_0^{i/10} y_1(t) \cdot \sin(i/10 - t) dt\right)^2$$

$$= \min_{c_k} \sum_{i=0}^{40} \left(-y_1(i/10) + i/10 + \sum_{k=-10}^{10} c_k \int_0^{i/10} \phi_{1,k}(t) \cdot \sin(i/10 - t) dt\right)^2 .$$

Here we can use more collocation points and we got better results. We now see the curve of the difference $y_1 - y$:

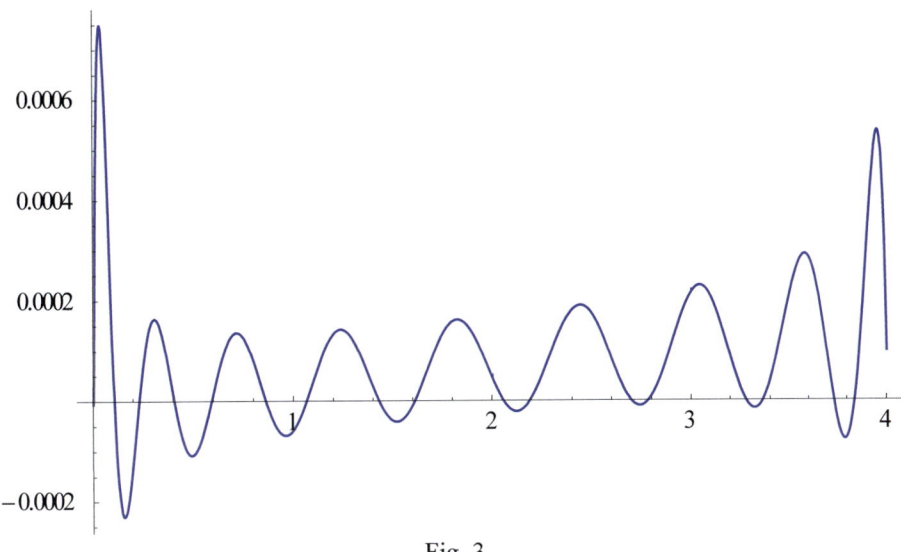

Fig. 3

With $k_{max} = -k_{min} = 15$ we get better results, too. Here is the graph of $y_1 - y$ in that case:

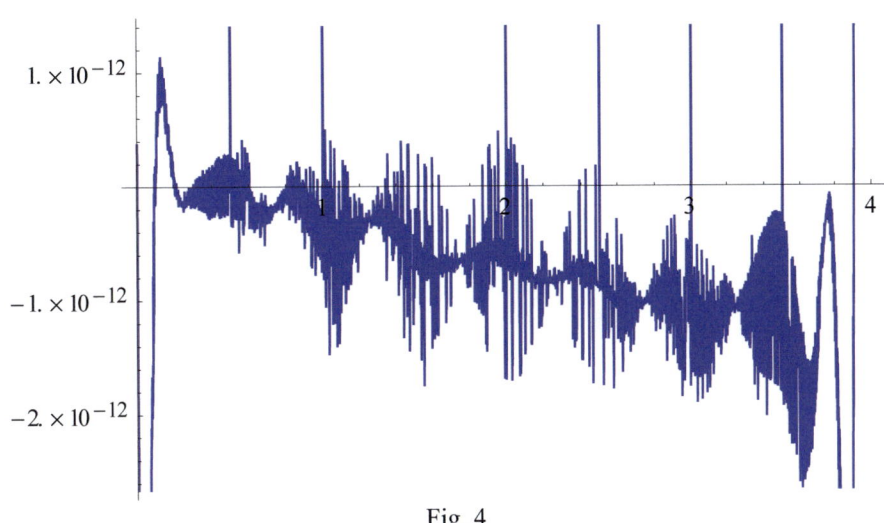

Fig. 4

**References**

[1] Schuchmann, M. (2012). "Approximation and Collocation with Wavelets. Approximations and Numerical Solving of ODEs, PDEs and IEs," *Osnabrück: DAV*

[2] Schuchmann, M.; Rasguljajew, M. (2013). Error Estimation of an Approximation in a Wavelet Collocation Method. *Journal of Applied Computer Science & Mathematics, No. 14 (7) / 2013, Suceava*

[3] Schuchmann, M.; Rasguljajew, M. (2013). Parameter Identification with a Wavelet Collocation Method in a Partial Differential Equation. *Journal of Approximation Theory and Applied Mathematics (JATAM) Vol. 1*

[4] Schuchmann, M.; Rasguljajew, M. (2013). An Approach for a Parameter Estimation with a Wavelet Collocation Method. *Journal of Approximation Theory and Applied Mathematics (JATAM) Vol. 1*

# Approximation of Non $L^2(R)$ Functions on a Compact Interval with a Wavelet Base

M. Schuchmann and M. Rasguljajew from the Darmstadt University of Applied Sciences

## Abstract

Comparing the approximations of functions on a compact interval $I$, we noticed that when $y$ is not in $L^2(R)$ the discrete least square method was significantly better than projecting $1_I y$ orthogonal to $V_j$ (with the indicator function $1_I$). This method even gives very good approximations when using relatively few basis elements. In this paper we show that the orthogonal projection from $1_I y$ to $V_j$ with the Shannon wavelet leads approximately to a Fourier series with a Gibbs effect. Because we cut a piece of $y$ with $1_I y$ it has consequences in the Fourier space so that even if $y$ is in $L^2(R)$ and its Fourier transform would have a good decay behavior, $1_I y$ has than a worse decay. But if we continue $1_I y$ so, that the continuation of y is in $L^2(R)$, we get a good approximation of $y$ on $I$ with the orthogonal projection.

## Introduction

In the wavelet theory a scaling function $\phi$ is used, which belongs to a MSA (multi scale analysis). From the MSA we know, that we can construct an orthonormal basis of a closed subspace $V_j$, where $V_j$ belongs to a the sequence of subspaces with the following property:

$$... \subset V_{-1} \subset V_0 \subset V_1 \subset ... \subset L^2(R),$$

$\{\phi_{j,k}(t)\}_{k \in Z}$ is an orthonormal basis of $V_j$ with $\phi_{j,k}(t) = 2^{j/2} \phi(2^j t - k)$.

We want to get an approximation of a function $y$ on a compact interval $I$. So $y$ must not be in $L^2(R)$ but only in $L^2(I)$.

We use the approximation function

$$y_j(t) := \sum_{k=k_{min}}^{k_{max}} c_k \cdot \phi_{j,k}(t) \ .$$

## The Approximation

In the following example we will see, that if we calculate the coefficients $c_k$ by the minimization of

(1) $$Q(c) = \sum_{i=0}^{m} (y_j(t_i) - y(t_i))^2$$

with $t_i \in I$ it will lead to much better results than if we calculate the coefficients $c_k$ with the orthogonal projection from $1_I y$ on $V_j$:

(2) $$c_k = \langle 1_I y, \phi_{j,k} \rangle = \int_I y(t) \cdot \phi_{j,k}(t) dt$$

Here $c_k$ depends on $j$ too, but for easier notation we write short $c_k$. This is an analogous result to [1], where we used an ODE and we minimized the residuals instead of (1). A reason for the worse approximation is, that a function with compact support like $1_I y$ is not very concentrated on special frequency areas in the Fourier space, that means the Fourier transform of $1_I y$ has not a good decay behaviour. For example $1_{[-1,1]}$ has the Fourier transform

$$Y(\omega) = \frac{1}{\sqrt{2\pi}} \int_{-1}^{1} e^{-i\omega t} dt = \sqrt{\frac{2}{\pi}} \cdot \frac{\sin(\omega)}{\omega} \ .$$

Because of the multiplication theorem the Fourier transform of $1_{[-1,1]} \cdot ^n$ (with $n \in N$) are derivatives of the $Y$ function multiplied with $i^n$ and have the same decay behaviour as $Y$. And generally, $1_I y$ has jumps at the points at the edges of the interval $I$, which leads to a poorer decay behaviour of the Fourier transform (see remarks). If we continue $1_I y$, so that the completion $\tilde{y}$ is quadratic integrabel on $R$, we get a much better approximation if we calculate the orthogonal projection form that completion $\tilde{y}$ on $V_j$. Here we use the function

$$h(t) = e^{-t^2}$$

and set (for $I = [a, b]$):

(3)
$$\tilde{y}(t) = \begin{cases} h(t-a) \cdot y(a) & \text{if } t < a \\ y(t) & \text{if } a \leq t \leq b \\ h(t-b) \cdot y(b) & \text{if } t > b \end{cases}$$

We can show, that $1_I y$ on $V_j$ can be approximated with a partial sum of a Fourier series, if we use the Shannon wavelet. The reason is, that if we calculate the inverse Fourier transform of $1_I y$ we get a function $g = \mathcal{F}^{-1}(1_I y)$, which has a compact support. Here we swap original space with Fourier space for the direct application of the Shannon theorem. Because of the Shannon theorem we know, that if $I \subseteq [-2^r \pi, 2^r \pi]$ than $g \in V_r$ and

(4)
$$g(s) = \sum_{k=-\infty}^{\infty} g_k \cdot \phi_{r,k}(s) \quad \text{with} \quad g_k = 2^{-r/2} g(k/2^r)$$

for almost all $s$. So the coefficients $g_k$ can be written as function values.

If we calculate the Fourier transform of $g$ we get

(5)
$$G(t) = \frac{2^{-r/2}}{\sqrt{2\pi}} \cdot \sum_k g_k \cdot 1_{[-2^r \pi, 2^r \pi]}(t) \cdot e^{-itk/2^r} \ .$$

This is a Fourier series of $1_I y$ (with respect to the interval $[-2^r \pi, 2^r \pi]$) and

(6)
$$G_{n_{min}, n_{max}}(t) = \frac{2^{-r/2}}{\sqrt{2\pi}} \cdot \sum_{k=n_{min},\ldots,n_{max}} g_k \cdot 1_{[-2^r \pi, 2^r \pi]}(t) \cdot e^{-itk/2^r}$$

is the approximation. Now we need a big summation area $[n_{min}, n_{max}]$ to get with $y_j$ a good approximation of $1_I \cdot y$, if $g$ has a worse decay behaviour and so a big $j$, because $\mathrm{supp}(Y_j) = [-2^j \cdot \pi, 2^j \cdot \pi]$. Thus with respect to $y_j$ we can only considers the $g_k$ with $|k|/2^r \in [-2^j \cdot \pi, 2^j \cdot \pi]$ and $n_{min}$ is the smallest integer $n$ with $n \geq -2^{j+r} \cdot \pi$ and $n_{max}$ the biggest integer $n$ with $n \leq 2^{j+r} \cdot \pi$, if $G_{n_{min},n_{max}}$ should be in $V_j$. Then, with growing $r$, the function $G_{n_{min},n_{max}}$ tends to the Fourier transform of $g \cdot 1_{[-2^j\pi, 2^j\pi]}$, i.e. to $y_j$. This is the reason for a worse approximation of $1_I \cdot y$ with an orthogonal projection on $V_j$ with a small $j$.

**Example:**

In this example we use the Shannon wavelet. We want to approximate $y(t) = e^{-t}$ on $I = [0, 1]$. The orthogonal projection from $1_I \cdot y$ on $V_3$ leads to a worse approximation, what we can see on the graph of $y_3$ ($-k_{min} = k_{max} = 24$):

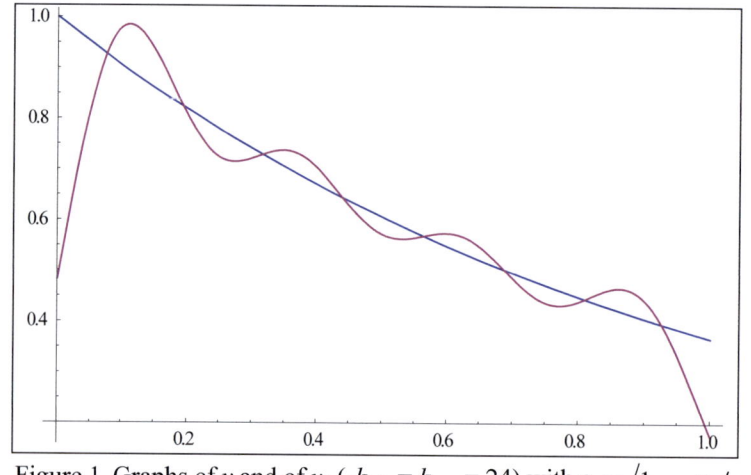

Figure 1. Graphs of $y$ and of $y_3$ ($-k_{min} = k_{max} = 24$) with $c_k = \langle 1_{[0,1]} y, \phi_{j,k} \rangle$

$$\|y - y_3\|_{L^2([0,1])} \approx 0.0881117$$
$$\|y - y_3\|_\infty \approx 0.517282 \text{ (on } I\text{)}$$

In the next figure we use instead of $y(t) = e^{-t}$ the function $m(t) = e^{-|t|}$ and if we calculate the orthogonal projection from $1_{[-3,3]} \cdot m$ on $V_3$ we get (we set $-k_{min} = k_{max} = 24$):

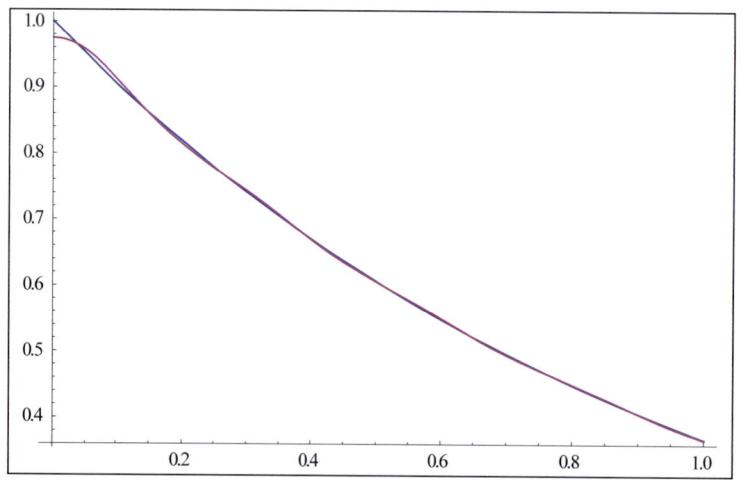

Figure 2. Graphs of $y$ and of $y_3$ ($-k_{min} = k_{max} = 24$) with $c_k = \langle 1_{[-3,3]} m, \phi_{j,k} \rangle$

$$\|y - y_3\|_{L^2([0,1])} \approx 0.00369947$$

$$\|y - y_3\|_\infty \approx 0.0257361 \text{ (on } I\text{)}$$

If we continue $I_1 y$ like in (3) with $a = 0$ and $b = 1$ then we get the following graph of $\tilde{y}$:

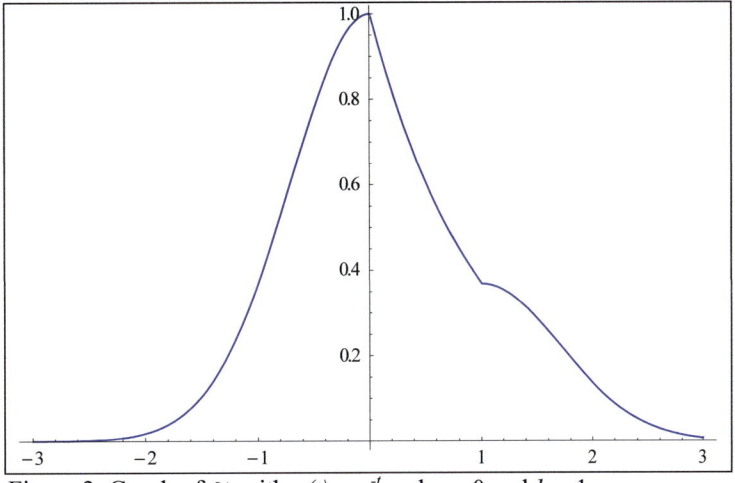

**Figure 3.** Graph of $\tilde{y}$ with $y(t) = e^{-t}$ and $a = 0$ and $b = 1$

Now we calculated the orthogonal $y_3$ projection of $1_{[-3,3]} \cdot \tilde{y}$ on $V_3$ (we can also project $\tilde{y}$ on $V_3$, but in the practical case we integrate only over a finite interval). Here we see the graph of $\tilde{y}$ and $y_3$ (with $-k_{min} = k_{max} = 24$):

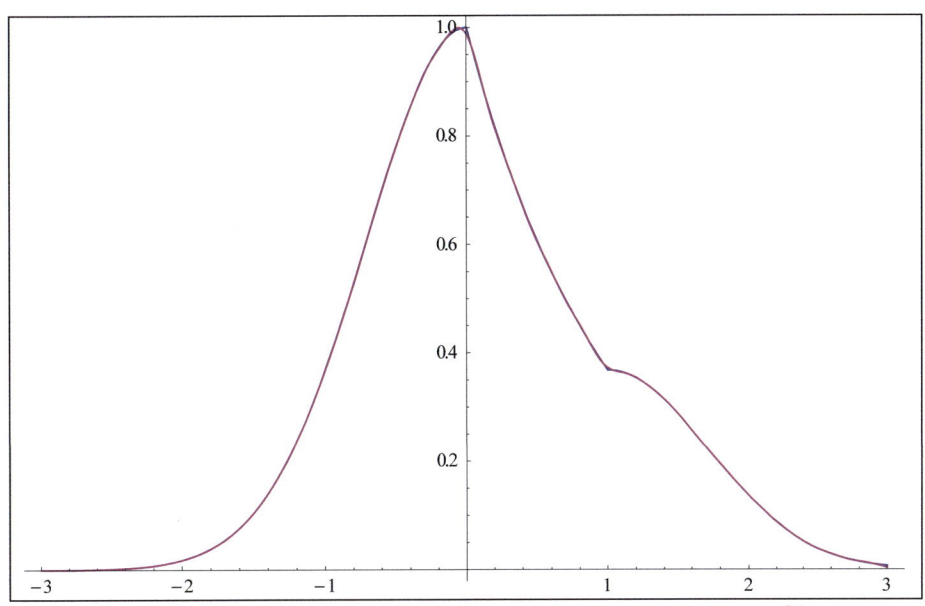

**Figure 4.** Graphs of $\tilde{y}$ and the orthogonal projection from $1_{[-3,3]} \cdot \tilde{y}$ on $V_3$

$$\|y - y_3\|_{L^2([0,1])} \approx 0.0020488$$

$$\|y - y_3\|_\infty \approx 0.0126504 \text{ (on } I\text{)}$$

With a differentiable $\tilde{y}$ we could get with a smaler $j$ even better results.

The graph of the amplitude spectrum of $1_{[0,1]} y$ (the absolute values of the Fourier transform of $1_{[0,1]} y$):

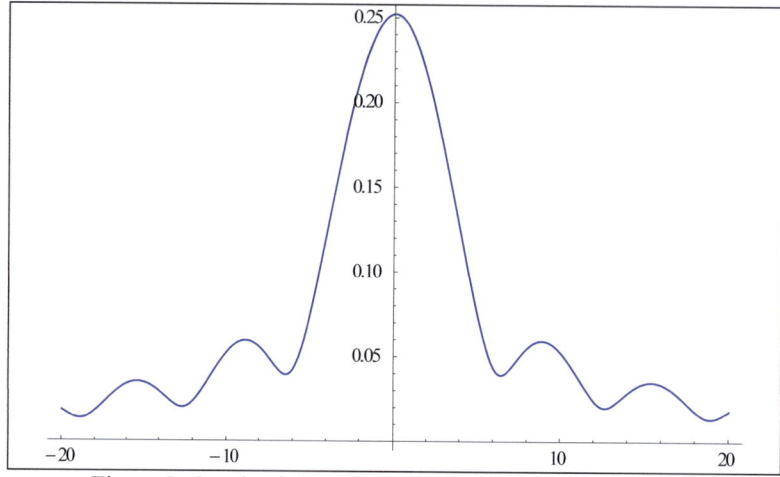

Figure 5. Graph of the amplitude spectrum of $1_{[0,1]} y$

And here the graph of the amplitude spectrum of $1_{[-3,3]} \cdot \tilde{y}$ (the absolute values of the Fourier transform of $1_{[-3,3]} \cdot \tilde{y}$):

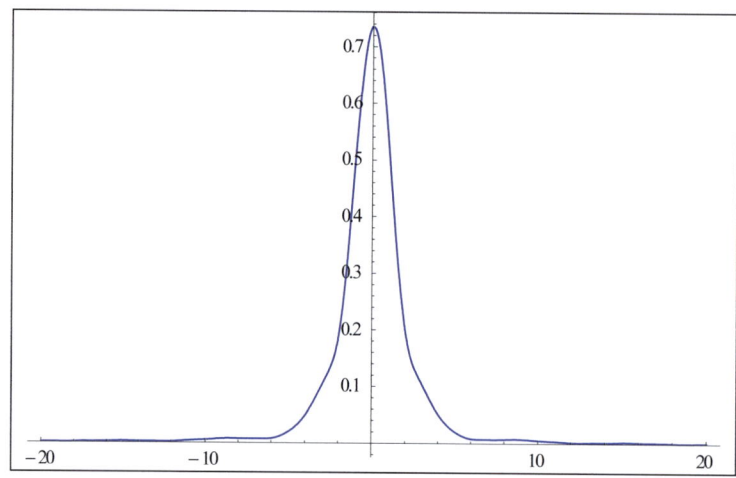

Figure 6. Graph of the amplitude spectrum of $1_{[-3,3]} \cdot \tilde{y}$

We can see that Fourier transform of $1_{[-3,3]} \cdot \tilde{y}$ has a much better decay behaviour as $1_{[0,1]} y$ which is the reason for a better approximation if we use the orthogonal projection on $V_j$ with a small $j$ as an approximation function.

Here is the graph of abs(g) ($r = 0$, because $[0,1] \subseteq [-2^r \cdot \pi, 2^r \cdot \pi]$, we could even set $r = -1$) for $1_{[0,1]} y$ ($g = \mathcal{F}^{-1}(1_1 y)$ and the band limited function $g$ can be written as a linear combination of bases elements of $V_0$, see (4)).

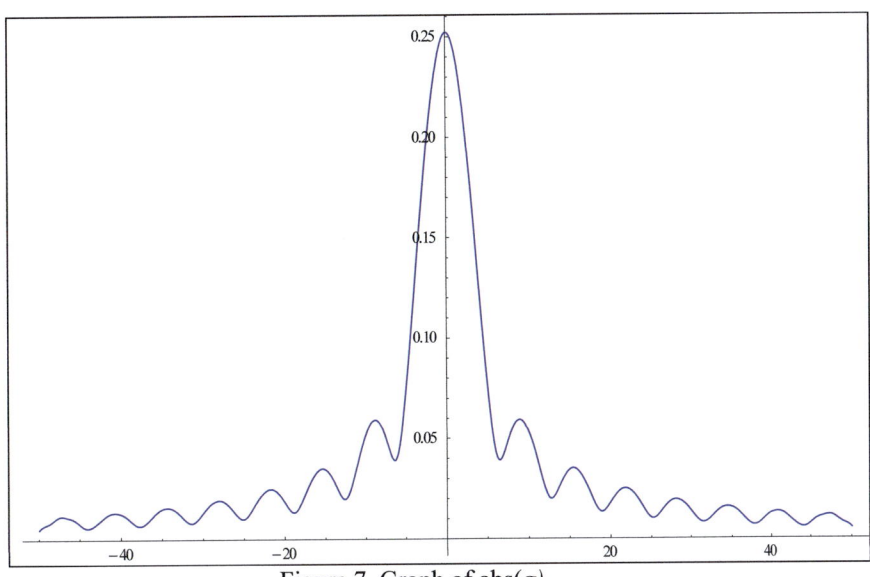
Figure 7. Graph of abs(g)

The Fourier transform of $g$ is a Fourier series of $1_{[0,1]}y$. Here is the graph of $G_{n_{min},n_{max}}$ (see (6)) and $1_{[0,1]}y$ for $-n_{min} = n_{max} = 50$:

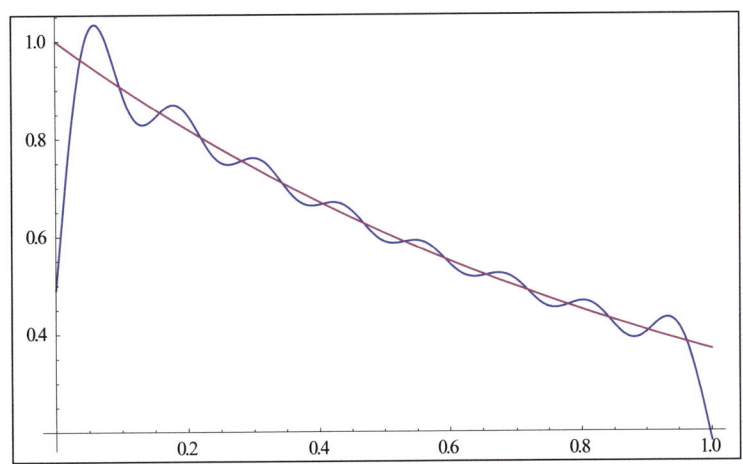
Figure 8. Graphs $1_{[0,1]}y$ and $G_{-50,50}$

$G_{-50,50}$ is a good approximation of the orthogonal projection from $1_{[0,1]}y$ on $V_4$ (because $-n_{min} = n_{max}$ is near $2^4 \cdot \pi = 2^{r+j} \cdot \pi$), although $r$ is not very big. Here is the graph of $G_{-100,100}$ and $1_{[0,1]}y$:

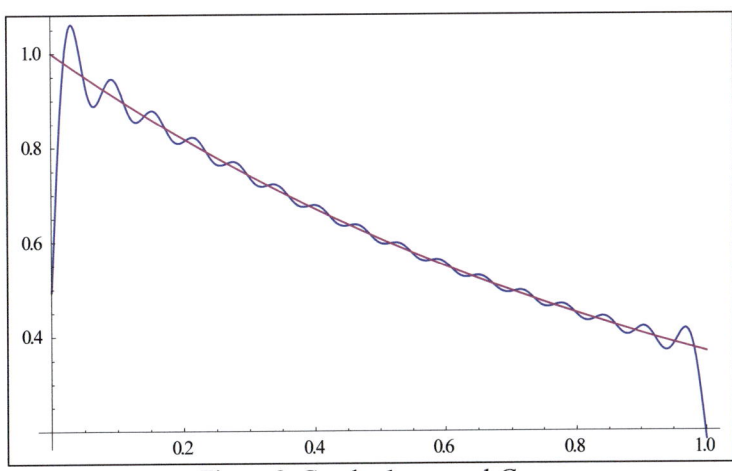
Figure 9. Graphs $1_{[0,1]}y$ and $G_{-100,100}$

Above we see the Gibbs effect and we see that we need a big summation area.

Here is the scheme of the transformation and projection:

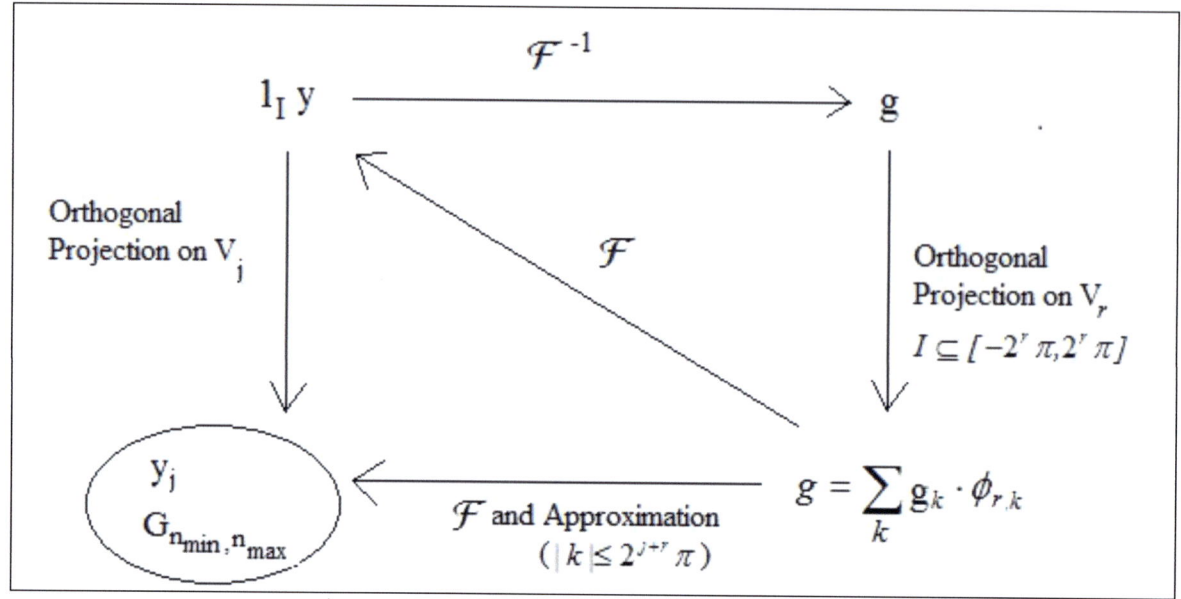

Figure 10. Scheme of the transformation and projection

At least we calculate a discrete approximation of y on $I = [0, 1]$. Here we set $-k_{min} = k_{max} = 10$ and we minimize Q (see (1)) with $j = 0$. We use $m = 20$ and $t_i = i/20$. That leads to a very good approximation.

Here are the graphs of $y$ and $y_0$:

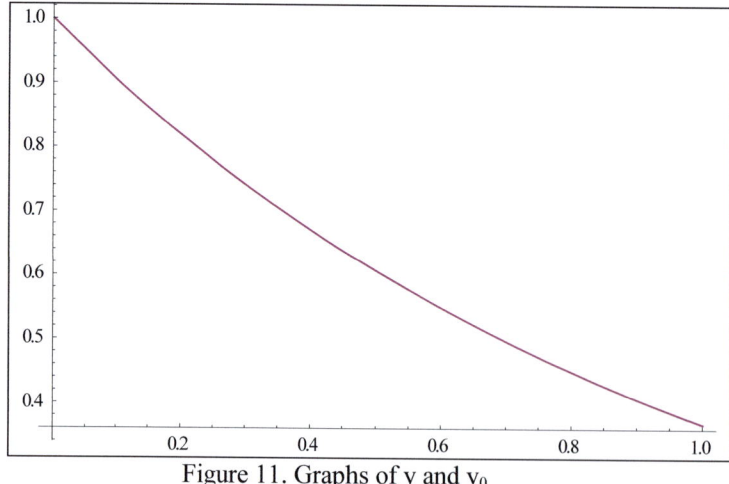

Figure 11. Graphs of y and $y_0$

Here is the graph of the error $y_0 - y$:

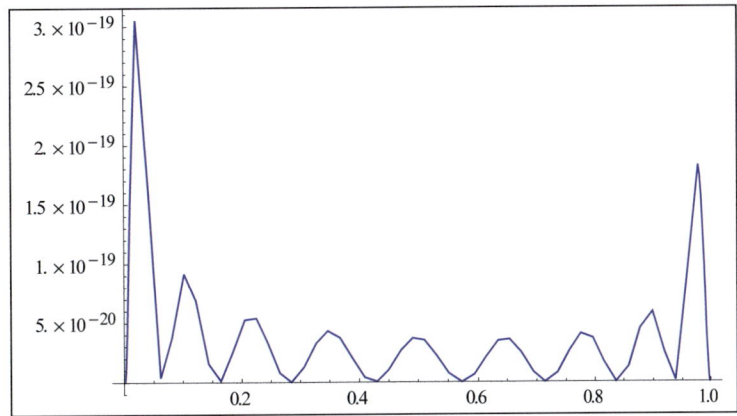

Figure 12. Graph of the error $y_0 - y$

**Remarks:**

1) Some authors suggest a regression, where the coefficients $c_k$ are calculated over the scalar product $c_k = \langle \tilde{y}, \phi_{j,k} \rangle$ with a help function $\tilde{y}$ through the points $(t_i, y_i)$ (with $i = 1, 2, ..., n$ and $t_i \in [a, b]$) and with supp $\tilde{y} = [a, b]$. Because generally the help function is not continuous we have the same problem of bad decay behaviour of the coefficients $c_k$ and we also need a big $j$.

2) If y has a jump at the point $t = \xi$ then we get for the Fourier transform of y (if $y' \in \mathcal{L}^1(R)$):

$$Y(\omega) = \frac{1}{\sqrt{2\pi}} \int_{-\infty}^{\infty} y(t) \cdot e^{-i\cdot\omega\cdot t} dt = \frac{1}{\sqrt{2\pi}} \int_{-\infty}^{\xi^-} y(t) \cdot e^{-i\cdot\omega\cdot t} dt + \frac{1}{\sqrt{2\pi}} \int_{\xi^+}^{\infty} y(t) \cdot e^{-i\cdot\omega\cdot t} dt$$

$$= \frac{1}{\sqrt{2\pi}} \left( \left[ y(t) \cdot \frac{1}{-i\cdot\omega} e^{-i\cdot\omega\cdot t} \right]_{-\infty}^{\xi^-} + \int_{-\infty}^{\xi^-} y'(t) \cdot \frac{1}{i\cdot\omega} e^{-i\cdot\omega\cdot t} dt + \left[ y(t) \cdot \frac{1}{-i\cdot\omega} e^{-i\cdot\omega\cdot t} \right]_{\xi^+}^{\infty} + \int_{\xi^+}^{\infty} y'(t) \cdot \frac{1}{i\cdot\omega} e^{-i\cdot\omega\cdot t} dt \right)$$

$$= \frac{1}{\sqrt{2\pi}} \left( \frac{y(\xi^+) - y(\xi^-)}{i\cdot\omega} e^{-i\cdot\omega\cdot\xi} + \int_{R\setminus\{\xi\}} y'(t) \cdot \frac{1}{i\cdot\omega} e^{-i\cdot\omega\cdot t} dt \right)$$

$$= \frac{1}{\sqrt{2\pi}} \left( O\!\left(\frac{1}{\omega}\right) + \int_{R\setminus\{\xi\}} y'(t) \cdot \frac{1}{i\cdot\omega} e^{-i\cdot\omega\cdot t} dt \right)$$

**References**

[1] M. Schuchmann, M. Rasguljajew. *An Approximation on a Compact Interval Calculated with a Wavelet Collocation Method can Lead to Much Better Results than other Methods.* Journal of Approximation Theory and Applied Mathematics (2013, Vol. 1)

[2] M. Schuchmann. *Approximation and Collocation with Wavelets. Approximations and Numerical Solving of ODEs, PDEs and IEs.* Osnabrück: DAV, (2012).

# Comparing Approximations of a Wavelet Collocation Method of Various Wavelets

M. Schuchmann and M. Rasguljajew from the Darmstadt University of Applied Sciences

## Abstract

In this paper we describe the application of a wavelet collocation method on different ODE's. Here we compare the approximation error of various Wavelets. The Shannon wavelet and the Meyer wavelet provides very good results. This method can be extended to unstable and stiff differential equations. In this work we also show how to set the Parameters of the collocation method and we present a general algorithm for this method.

## Introduction

As part of a research project we investigated how to determine the optimal parameters for a wavelet collocation method. In the classical approach to collocation methods the approximation function is based on polynomials. These methods are equivalent to implicit Runge-Kutta method, which are used in stiff problems and boundary value problems. In the wavelet collocation method the approximation functions are constructed by a wavelet base.

There are a lot of different parameters to be set which brings up the question how useful the approximation function is if the exact solution is unknown. Here, we performed a series of simulations in which a criterion was found which theoretical could be used for a estimation. Using regression analysis, there were significant correlations between this criterion and the mean square approximation error. The criterion for evaluating the approximation of the herein described wavelet collocation method was used and theoretically justified by an estimate in [16] by M. Schuchmann.

In this study, various wavelets were compared, because there are whole families of wavelets, such as Daubechies wavelets, the Meyer wavelets or the Battle Lemarié wavelets available. One wavelet that does not have compact support, and not even a high order, provided very good approximations. The approximation functions can even be used to for extrapolations. This wavelet was the Shannon wavelet, which is infinitely differentiable and, unlike many other wavelets has a mother wavelet and a scaling function (also called father wavelet) which can be written in closed form.

We will use an approach in which the trial function is composed of a wavelet basis. Instead of solving a system of equations and to set the residuals equal to zero at certain points, we minimize the sum of squared residuals (at the collocation points), so that we are not restricted in the number of collocation points.

The advantage of the wavelet collocation method is that like other collocation method it also can be applied to stiff differential equations. Moreover, it can even be used in non-stable problems (see [15]). As an approximation we not only get points but an approximation function. Compared to other collocation method, for example, based on polynomials (see [3]), one can also cover a larger interval with an approximation, i.e. you do not have to use composite functions for small subintervals. These are the advantages of a wavelet collocation method as well as using the approximation for extrapolation outside the original approximation interval. As a disadvantage, you could argue that if there is a differential equation for which one needs no boundary value problem methods (i.e. if they are not stiff or unstable) more computing time may be needed since a minimization problem or a system of equations must be solved.

There are wavelets which are called interpolating wavelets with special properties. There are a number of publications on these wavelets. These deal with error estimates and also with the approximation of the solutions of initial value problems and boundary value problems (for ordinary and partial differential equations), see [23] and [4], as well as with the sinc collocation (see [5], [1], [10]) with special support points ("sinc grid points", see [10]). The scaling function of the Shannon wavelets which we use later are based on the sinc function and also have interpolating properties (see [18]).

In [9] a quasi-Shannon wavelet is used to approximately solve a boundary value problem (with second-order ordinary differential equation). The scaling function of the Shannon wavelets is weighted by a Gaussian function, so that the decay of the scaling function is improved.

Error estimates are provided for the Shannon wavelet (i.e. a sinc collocation with a transformation) in [10] and [1]. For interpolating wavelets estimates can be found in [19] and [6].

The method described here can also be applied to partial differential equations (see [17]). In addition, the method can also be used for parameter identification. For that an estimate in two steps was tested by us and M. Schuchmann has developed an error estimate, which was published by us in [18].

In the wavelet theory a scaling function $\phi$ is used, which belongs to a MSA (multi scale analysis). From the MSA we know, that we can construct an orthonormal basis of a closed subspace $V_j$, where $V_j$ belongs to a the sequence of subspaces with the following property:

$$... \subset V_{-1} \subset V_0 \subset V_1 \subset ... \subset L^2(R),$$

$\{\phi_{j,k}(t)\}_{k \in Z}$ is an orthonormal basis of $V_j$ with $\phi_{j,k}(t) = 2^{j/2} \phi(2^j t - k)$.

We use the following approximation function

$$y_j(t) := \sum_{k=k_{min}}^{k_{max}} c_k \cdot \phi_{j,k}(t) \quad , \text{with } \phi \in C^r(R).$$

$k_{max}$ and $k_{min}$ depend on the approximation interval $[t_0, t_{end}]$ (see [7]). $r$ is the order of the ODE.

Now we can approximate the solution of an initial value problem $y' = f(y,t)$ and $y(t_0) = y_0$ by minimization of the following function

$$(1) \quad Q(c) = \sum_{i=1}^{m} \left\| y_j'(t_i) - f(y_j(t_i), t_i) \right\|_2^2 + \left\| y_j(t_0) - y_0 \right\|_2^2 .$$

For $m = |k_{max} - k_{min}|$ we get an equivalent problem:

$$y_j'(t_i) = f(y_j(t_i), t_i) \text{ for } i = 1, 2, ...., m \text{ and } y_j(t_0) = y_0.$$

We will use equidistant points or collocation points $t_i$ with $t_i = t_0 + i \cdot h$ and

$$h = \frac{t_{end} - t_0}{m}.$$

To detect large residuals in other places as the collocation points, we have a further value used for comparison with $Q_{min}$ (here in $y_j$ the vector $c$ will be set to the value in the minimum of $Q$, see (1)).

$$Q_a = \sum_{i=1}^{m_a} \left\| y_j'(\tau_i) - f(y_j(\tau_i), \tau_i) \right\|_2^2 + \left\| y_j(t_0) - y_0 \right\|_2^2$$

with $\tau_i = t_0 + i \cdot h/a$. $m_a = a \cdot m$ with $a > 1$ as an integer. Since the wavelet collocation method provides a whole approximation function $y_j$ and not only points, we can calculate $Q_a$ without additional effort. If $Q_a \gg Q_{min}$ (and $Q_{min}$ was very small) then $m$ (the number of collocation points) should be increased. When comparing $Q_{min}$ with $Q_a$, $Q_a$ should be weighted by $1/a$ if $a$ is large. In the simulations $a = 2$ proved sufficient.

$Q_a$ can additionally be justified by an error estimation of the residuals at theoretically any number of points. This was derived by M. Schuchmann. In this error estimate a certain value occurs as a factor. $Q_a$ represents the Riemann sum for this value i.e. this can be approximated by $Q_a$. For this we use the following theorem:

**Theorem:**
Let $y' = f(y, t)$ with $y(t_0) = y_0$ and let (for $t \geq t_0$)

$$\|y_j'(t) - f(y_j(t), t)\| \leq M(t),$$

$$\|f(y(t), t) - f(y_j(t), t)\| \leq l(t) \cdot \|y(t) - y_j(t)\| \text{ with } l(t) > 0$$

and

$$y_j(t_0) = y_0.$$

With

$$L(t) = \int_{t_0}^{t} l(s) ds$$

follows (for $t_{end} \geq t_0$):

$$\left\| y_j(t_{end}) - y(t_{end}) \right\| \leq e^{L(t_{end})} \cdot \left\| e^{-L} \right\|_{L^2([t_0, t_{end}])} \cdot \left\| M \right\|_{L^2([t_0, t_{end}])}$$

The proof can be found in [16]. The factor on the right hand side of the inequality can now be approximated with $Q_a$ by

$$\|M\|_{L^2([t_0,t_{end}])} \approx \sqrt{\frac{t_{end}-t_0}{m_a}} Q_a$$

Analogous we could treat boundary conditions instead of the initial condition. This method can be even used analogous for PDEs, ODEs of higher order or ODEs, which have the Form $F(y',y,t) = 0$.

If we have a second Order ODE

$$F(y'',y',y,t) = 0$$

with boundary conditions

$$y(t_0) = y_0$$
$$\text{and}$$
$$y(t_{end}) = y_{end}$$

like in the following example, we minimize

$$Q(c) = \sum_{i=1}^{m} \|F(y_j''(t_i), y_j'(t_i), y_j(t_i), t_i)\|_2^2 + \|y_j(t_0) - y_0\|_2^2 + \|y_j(t_{end}) - y_{end}\|_2^2 .$$

Analogous we treat conditions of the form

$$y(t_0) = y_0$$
$$\text{and}$$
$$y'(t_0) = y'_0$$

## Comparing the Orthogonal Projection of y in $V_j$

Now we want to approximate two functions in the following two examples, which are not quadratic integrable on $R$.

**Example I:**
We begin with an approximation of the function $y(t) = e^{-t}$ on $I = [0, 2]$. $y$ is not in $L^2(R)$, but every on $I$ continuous function is in $L^2(I)$ or $1_I y$ (with indicator function $1_I$ of $I$) is in $L^2(R)$. So we set $k_{max} = -k_{min} = 20$ and we calculate an approximation function by an orthogonal projection from $1_I y$ on $V_3$. Therefore we calculate the coefficients of the approximation function over a scalar product (compare [v5]):

$$c_k = \langle 1_{[0,2]} y, \phi_{j,k} \rangle = \int_0^2 y(t) \cdot \phi_{1,k}(t) dt$$

With the Shannon wavelet we get a worse approximation (dashed line is thee graph of *y*). We consider in the graph only the interval [0.5,1] because of we cut the original function y at the edges we get worse approximation at the edges.

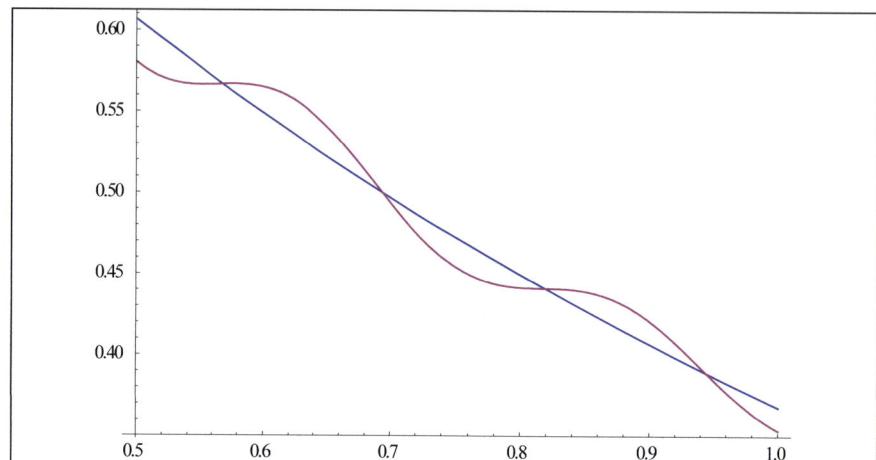

Figure 1. Graphs of $y_3$ (orthogonal projection form $I_3 y$ on $V_3$) and y

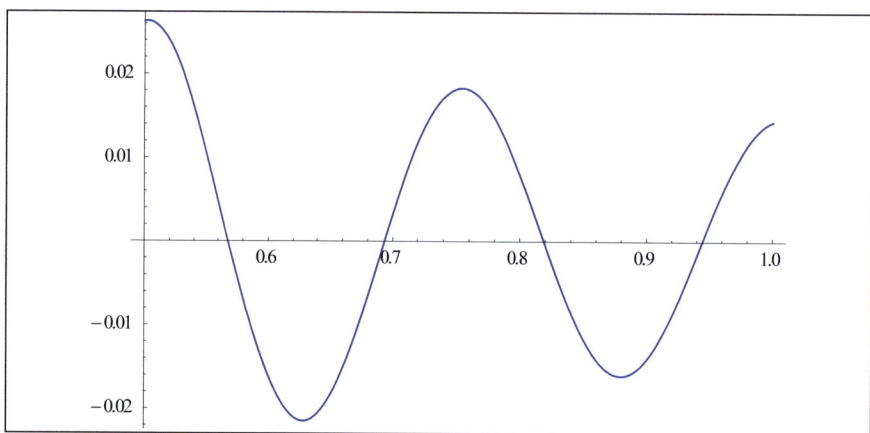

Figure 2. Graphs of $y - y_3$

With the Daubechies wavelet of order 8 we get no good approximation, but better approximation on $I = [0; 2]$, but in the midle of the interval $I$:

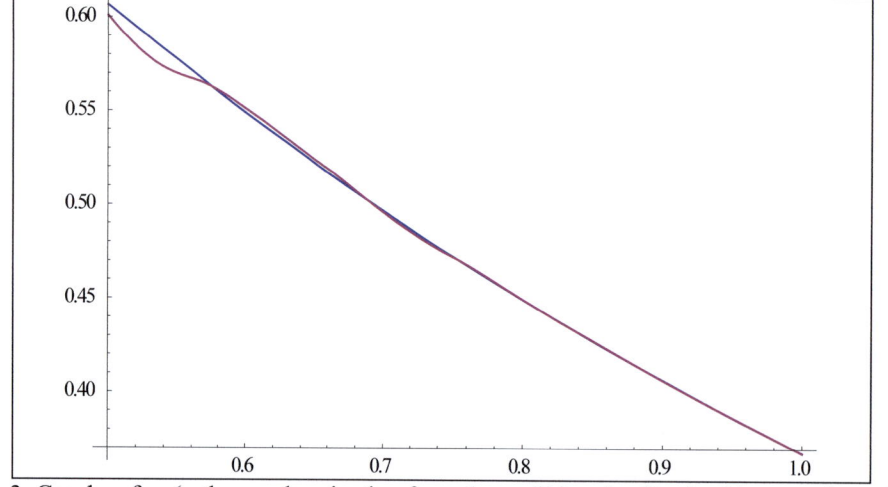

Figure 3. Graphs of $y_3$ (orthogonal projection form $I_3 y$ on $V_3$) and y, Daubechies wavelet order 8

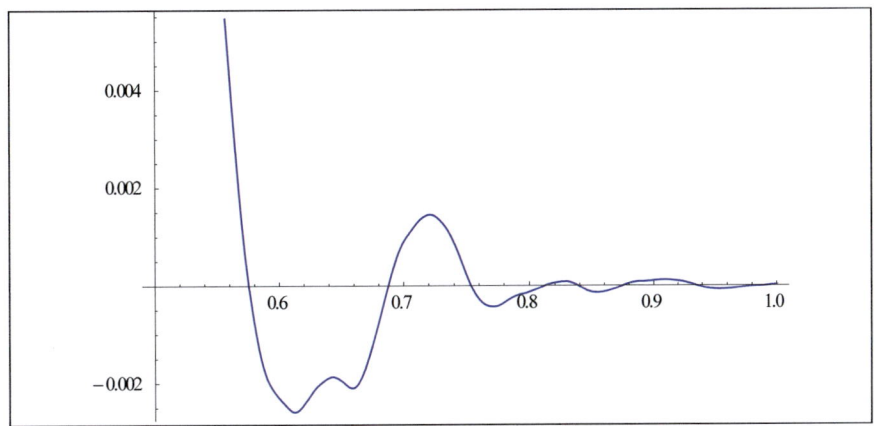

Figure 4. Graphs of y - $y_3$, Daubechies wavelet order 8

But the approximation with the wavelet collocation method can be much better with the Shannon wavelet, what we see in the following examples in many examples too.

Now we calculate the coefficients $c_k$ by the minimization of $Q$ (see (1)). We use the initial value problem $y' = -y$, $y(0) = 1$ and set even $j = 1$. We use the collocation points $t_i = i/20$ with $i = 0, 2, ..., 40$ and the Shannon wavelet.

Graphically we see no difference between the approximation function $y_1$ and $y$ on $I$:

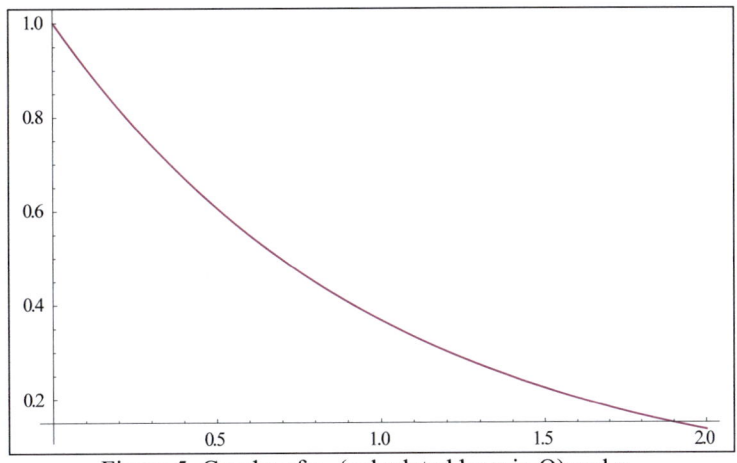

Figure 5. Graphs of $y_1$ (calculated by min Q) and y

Here is the graph of the difference function $y_j - y$:

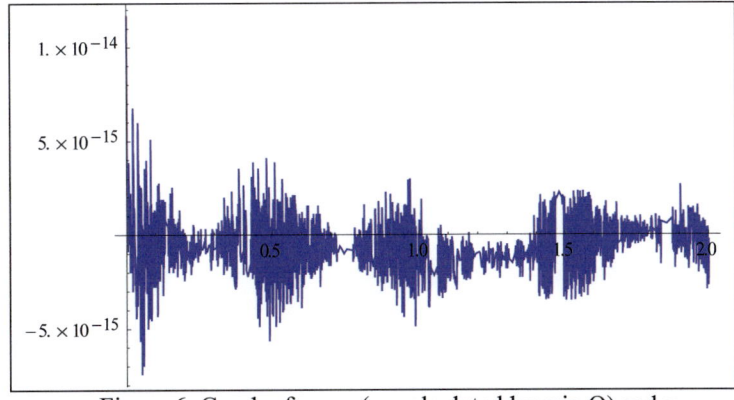

Figure 6. Graph of $y_1$ - y ($y_1$ calculated by min Q) and y

With the Daubechies wavelet of order 8 (D8) we get the following graph:

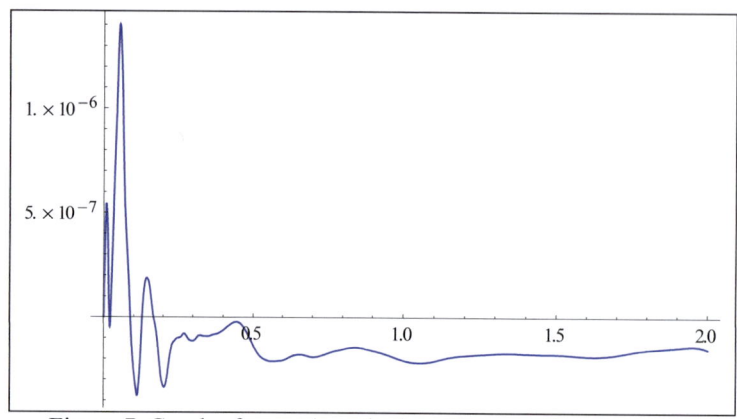

Figure 7. Graph of $y_1$ - y ($y_1$ calculated by min Q) and y, with D8

## Example II: Boundary value problem with a second order differential equation

Consider the following boundary value problem:

$$F(y'',y',y,t) = y'' - 1/\zeta \cdot (y - (\zeta \cdot \pi^2 + 1)\cos(\pi \cdot t)) = 0 \text{ with } y(-1) = 0 \text{ and } y(1) = 0, \zeta > 0.$$

This example was also used in the chapter with the title "comparing different wavelets" (Sample 8) and was found as test problem 14 on the website of Jeff Cash (Imperial College, London). If we write the problem as a first order system, with $y_1 = y'$ and $y_2 = y$, then

$$y_1' = 1/\zeta \cdot (y_2 - (\zeta \cdot \pi^2 + 1)\cos(\pi \cdot t))$$
$$y_2' = y_1$$

respectively $$\begin{pmatrix} y_1' \\ y_2' \end{pmatrix} = \underbrace{\begin{pmatrix} 0 & 1/\zeta \\ 1 & 0 \end{pmatrix}}_{=A} \cdot \begin{pmatrix} y_1 \\ y_2 \end{pmatrix} + \begin{pmatrix} -1/\zeta \cdot (\zeta \cdot \pi^2 + 1)\cos(\pi \cdot t) \\ 0 \end{pmatrix}.$$

The matrix A has the eigenvalues $\lambda_{1/2} = \pm \dfrac{1}{\sqrt{\zeta}}$.

Thus we see that at small $\zeta$ the solution function is composed of a function with a steep incline and a sharply decreasing function, which can lead to problems with numerical methods.

We are looking for an approximation on the interval $I = [-1, 1]$ and set $\zeta = 0.01$. We minimize

$$Q(c) = \sum_i (F(y_j''(t_i), y_j'(t_i), y_j(t_i), t_i))^2 + (y_j(-1) - 0)^2 + (y_j(1) - 0)^2$$

with the collocation points $t_i = i \cdot h$ (mit $i = 1, 2, \ldots, m$, $m = r \cdot k_{max}$), with $h = 2/(r \cdot k_{max})$ and $k_{min} = -k_{max}$. We use again $k_{max} = 15, 20, 25$, $r = 1, 2, 3$ and $j = 0, 1, 2$.

$Q_a$ is defined here in analogy to (5) for this boundary value problem:

$$Q_a = \sum_i (F(y_j''(\tau_i), y_j'(\tau_i), y_j(\tau_i), \tau_i))^2 + (y_j(-1) - 0)^2 + (y_j(1) - 0)^2$$

In the graphs the mean square error *mse* is again displayed with

$$mse = \frac{1}{101} \sum_{i=0}^{100} (y(2i/100 - 1) - y_j(2i/100 - 1))^2 \quad .$$

We start with the Shannon wavelet. Below are the graphs of the approximations that were relatively goodwith following triples $(j, k_{max}, m)$: $(0, 20, 40)$, $(1, 15, 30)$, $(1, 20, 40)$ and $(2, 25, 50)$. It is noticable that here $r = 2$ i.e. $m = 2 \cdot k_{max} = |k_{max} - k_{min}|$. Below are the graphs of $y_j$, $y$ and $y_j - y$.

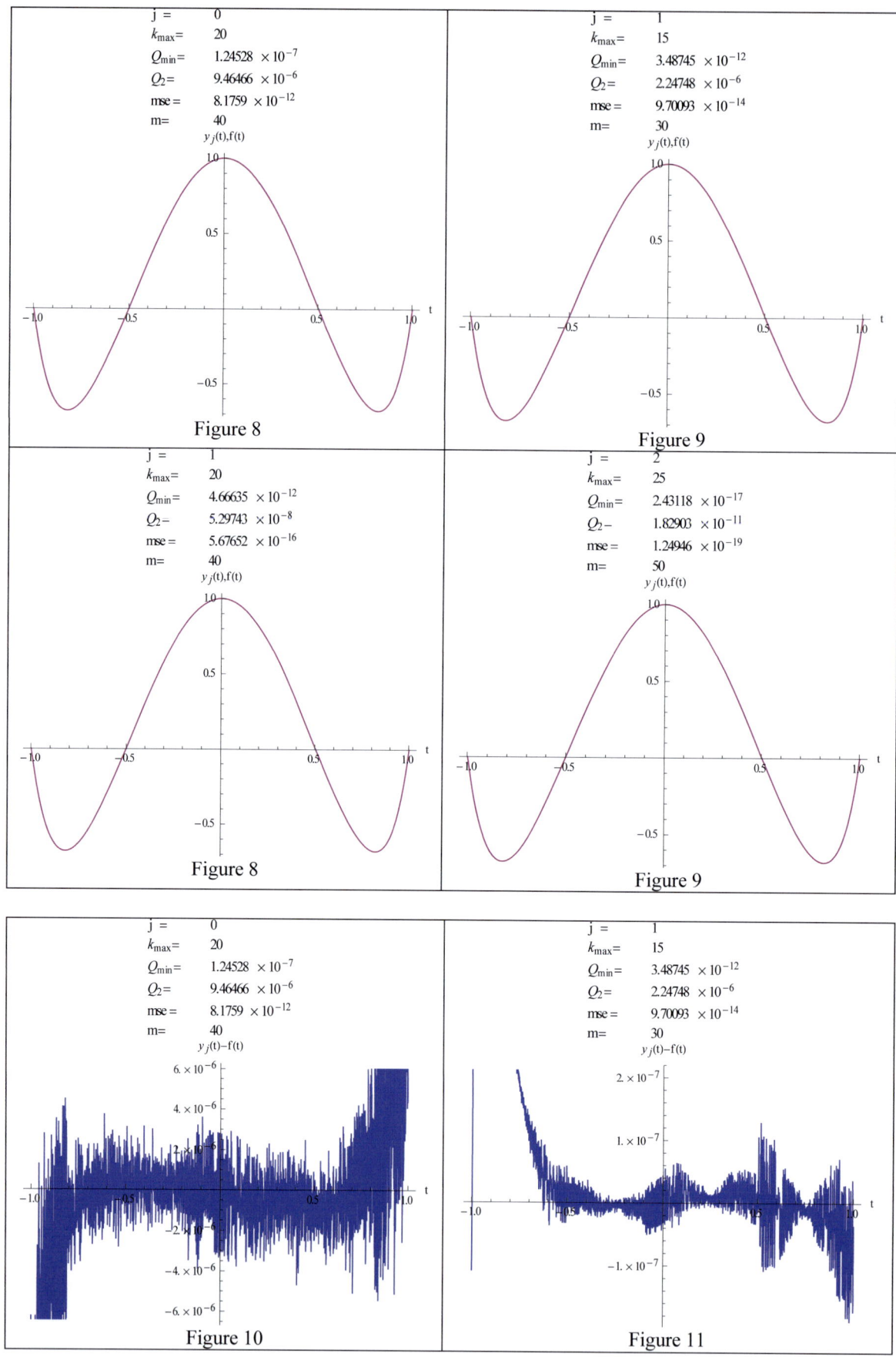

Figure 8

Figure 9

Figure 8

Figure 9

Figure 10

Figure 11

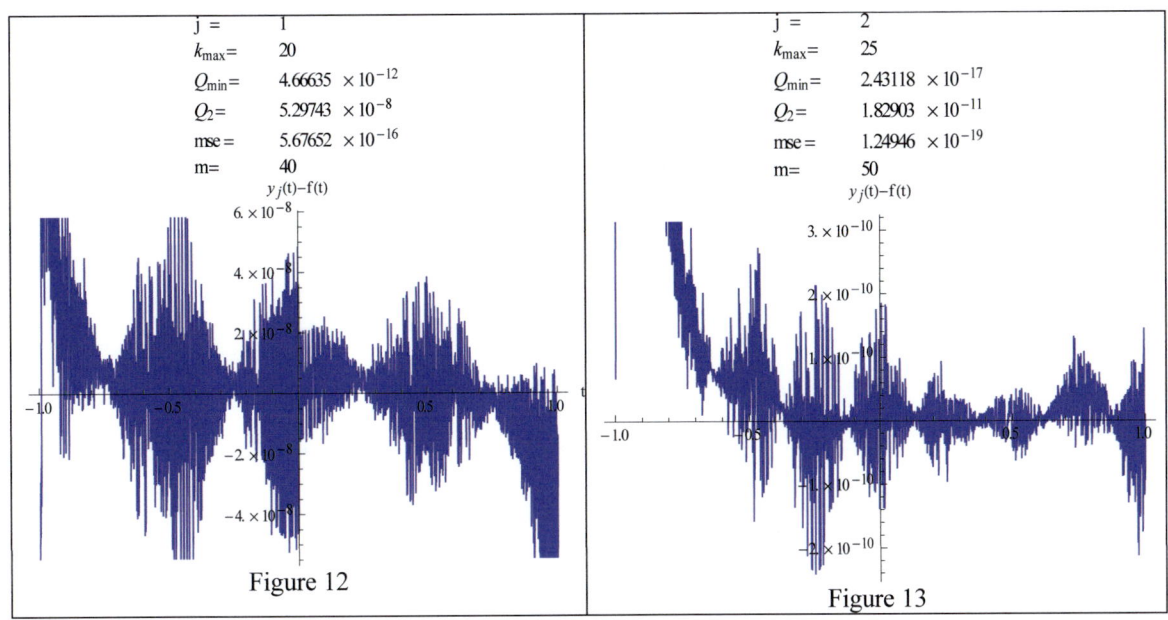

Figure 12

Figure 13

With the points $(-ln(Q_{min}), -ln(mse))$ (with various $j$, $k_{max}$ and $r$) a regression is calculated. Here again we see a relationship. However, more clearly the relationship is seen in the regression with the points $(-ln(Q_2), -ln(mse))$.

Regression with the points $(-ln(Q_{min}), -ln(mse))$:

|   | Estimate | SE | TStat | PValue | |
|---|----------|----------|---------|-------------------------|---|
| 1 | 6.14361  | 1.34266  | 4.57571 | 0.00011192              | , RSquared → 0.923205 |
| x | 1.03175  | 0.0595142| 17.3362 | $1.9112 \times 10^{-15}$ | |

Table 1

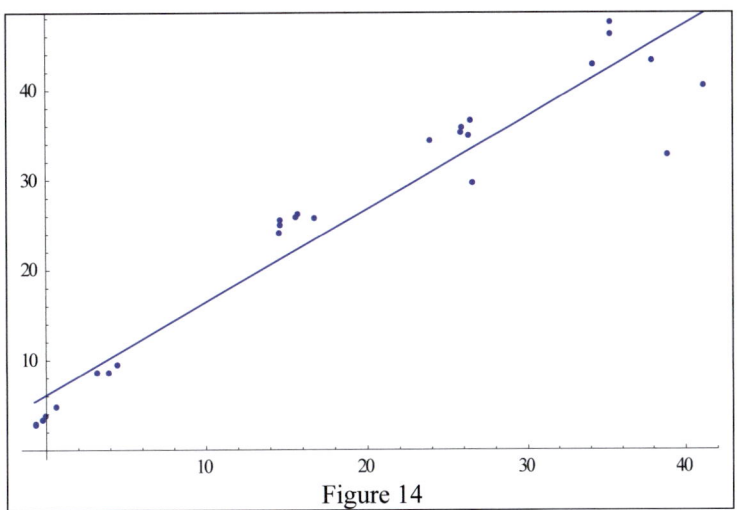

Figure 14

Regression with $(-ln(Q_2), -ln(mse))$:

|   | Estimate | SE | TStat | PValue | |
|---|----------|----------|---------|--------------------------|---|
| 1 | 8.12563  | 0.801252 | 10.1412 | $2.41445 \times 10^{-10}$ | , RSquared → 0.967851 |
| x | 1.41116  | 0.0514383| 27.434  | $3.50593 \times 10^{-20}$ | |

Table 2

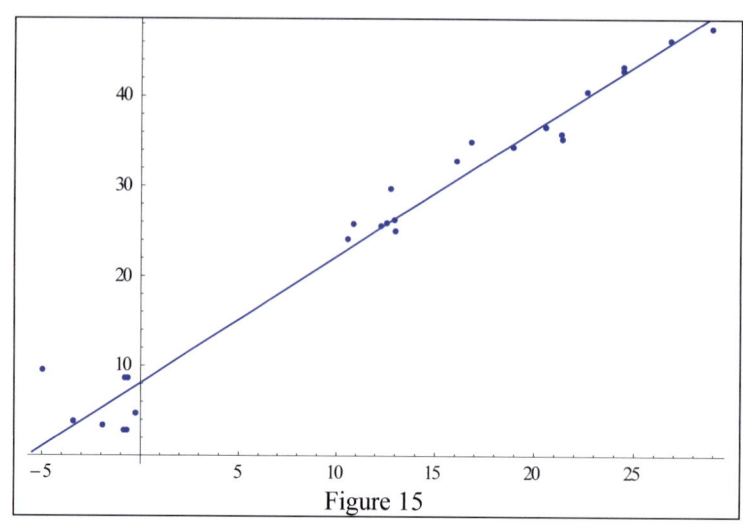
Figure 15

Now for comparison follows a regression of -ln($Q_2$) to -ln(mse) using the Meyer wavelet:

|   | Estimate | SE | TStat | PValue |   |
|---|---|---|---|---|---|
| 1 | 8.54055 | 0.747398 | 11.427 | $2.03499 \times 10^{-11}$ | , RSquared → 0.970825 |
| x | 1.45183 | 0.0503363 | 28.8426 | $1.04043 \times 10^{-20}$ |   |

Table 3

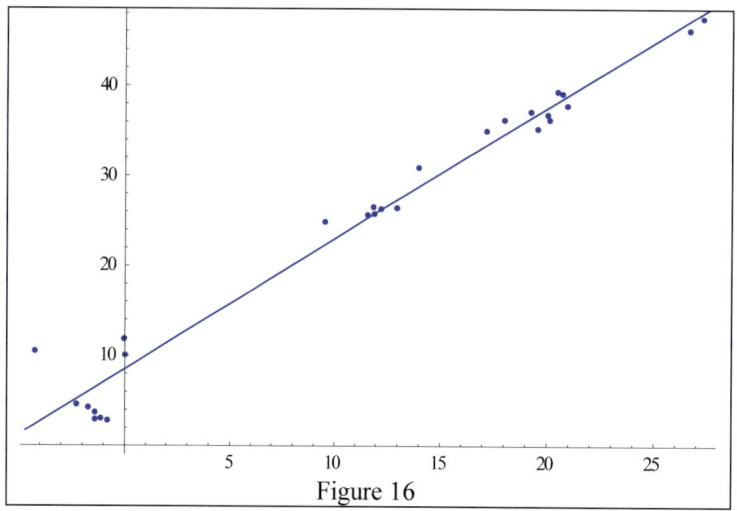
Figure 16

The Daubechies wavelet of order 8 as well as the Battle-Lemarié wavelet provided no useful approximations. This could be recognized by a very high $Q_{min}$ and $Q_2$ (see the chapter with the title "Comparison of different wavelets").

Even the NDSolve function of Mathematica 8 (also Mathematica 9) has problems with this boundary value problem.

There is a note displayed:
*NDSolve::berr: There are significant errors _{$4.85642 \times 10^{-29}$, $-1.83451 \times 10^{-6}$}_ in the boundary value residuals. Returning the best solution found.*

Here is the graph of the solution curve computed by NDSolve. At $t = 1$ we see major deviations:

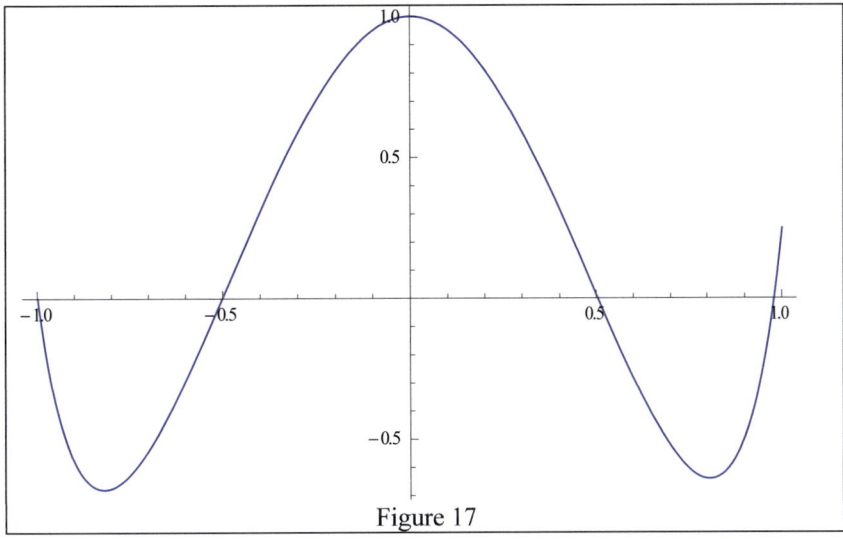
Figure 17

If for example $\zeta = 0.001$ is set, then the problems get even more severe:

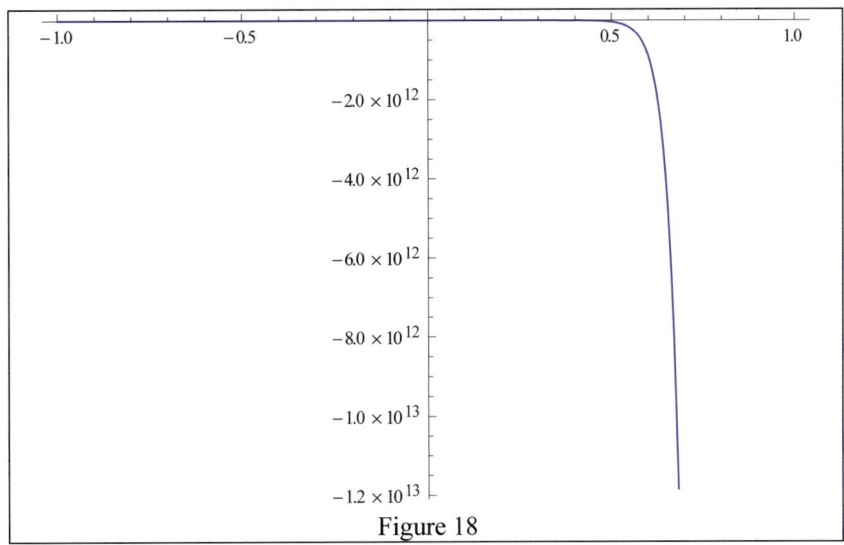
Figure 18

Mathematica displayed following note here:
*NDSolve::bvluc: The equations derived from the boundary conditions are numerically ill-conditioned. The boundary conditions may not be sufficient to uniquely define a solution. The computed solution may match the boundary conditions poorly.*
*NDSolve::berr: There are significant errors _{-1.10934×10$^{-30}$,-4.30118×10$^{11}$}_ in the boundary value residuals. Returning the best solution found.*

With this smaller $\zeta$ the wavelet collocation method has no problems, but it can cause big deviations between $y_j$ and $y$ in the neighbourhood of $t = -1$ and $t = 1$ (i.e. in the vicinity of the interval limits of the approximation interval $I$) provided $j$ is too small. This is due to the relatively large slope of $y$ in this area. For this reason the collocation points $t_i = i \cdot h$ should begin with $i = 0$, so that the slope at $t = -1$ is considered in $Q$.

With a smaller $j$ also relatively large values of $Q_2$ can occur; even if the whole approximation (or without the areas at the edge of approximation interval $I$) is good. This is due to the fact that $d(t) = (F(y_j''(t), y_j'(t), y_j(t), t))^2$ becomes relatively large in the aforementioned neighbourhood.

With such types of functions the points $\tau_i$ on the edge of the approximation area could be left out, if a good approximation on the inner part of the interval $I$ is needed and this approximation should be identified with $Q_2$.

Now we set $\zeta = 0.001$ and we minimize $Q$. Below are the graphs of $y_j$ and $y$ and the graph of $y_j - y$ for $j = 3$ and $r = 3$:

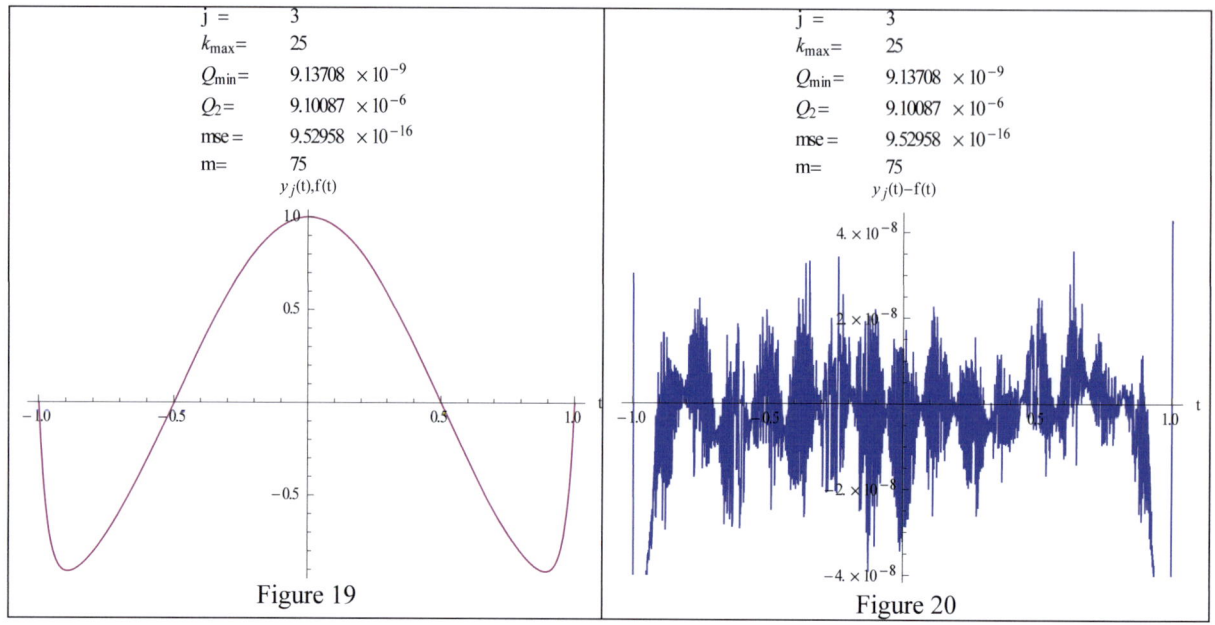

Figure 19                                       Figure 20

If $j$ is too small there are problems on the edges and $Q_{min}$ still relatively large. For example when $j = 1$ and $k_{max} = 15$:

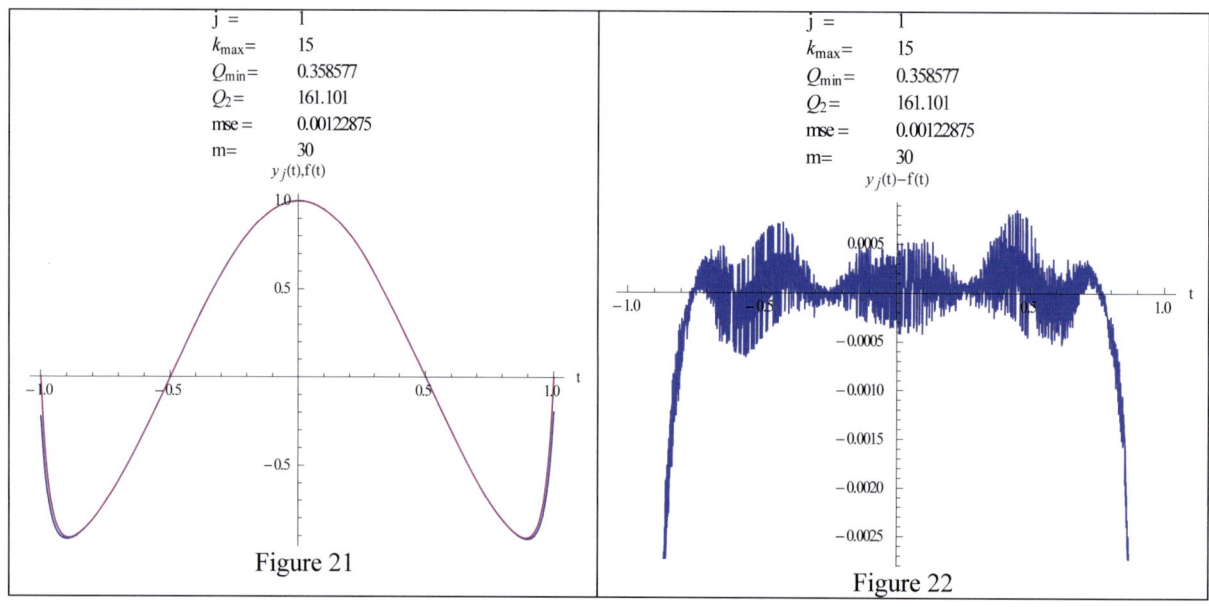

Figure 21                                       Figure 22

Or when $j = 1$ and $k_{max} = 25$:

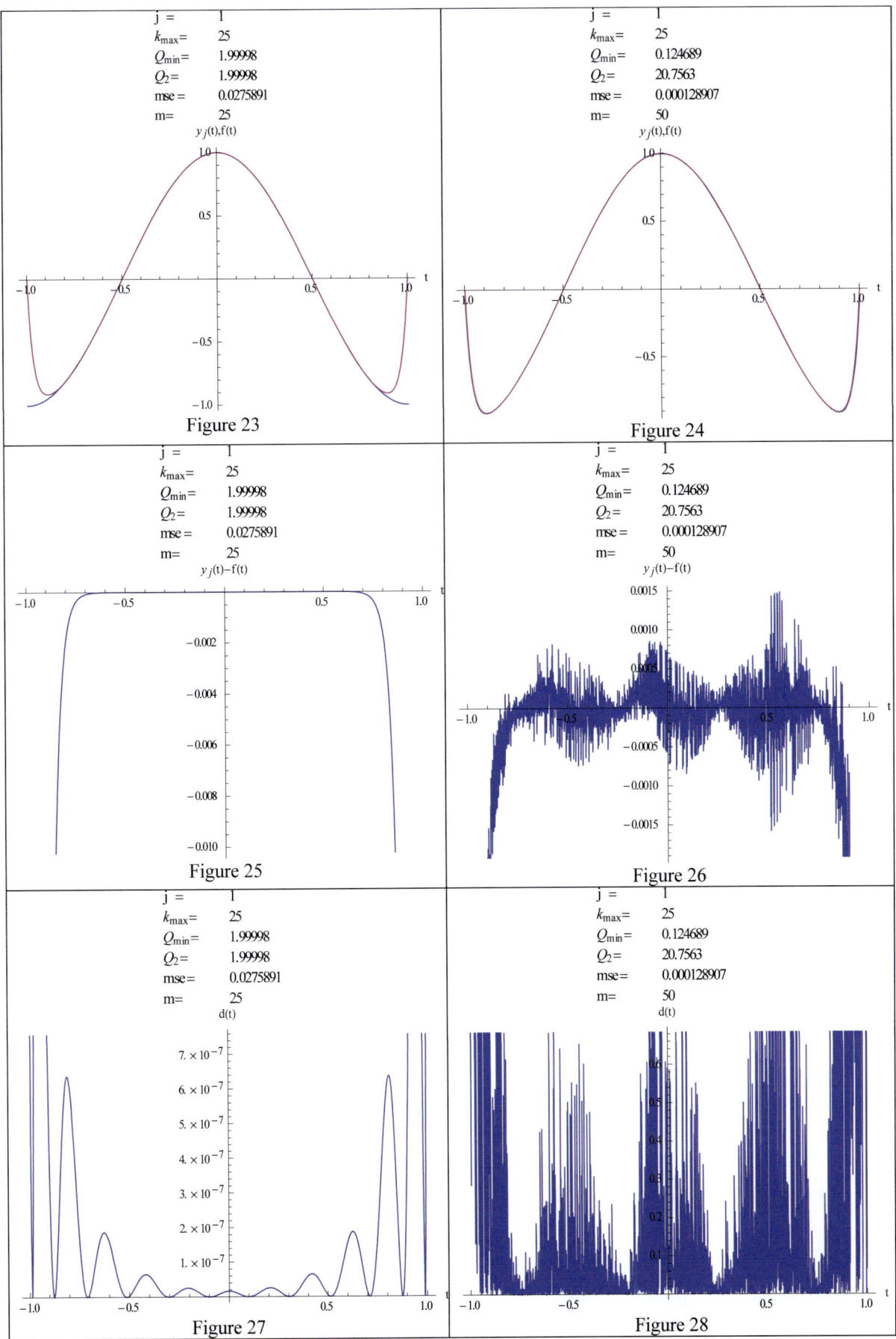

# The Algorithm

If no further information is available, one can start with $j = 1$ and $m = |k_{max} - k_{min}|$ and minimize $Q$. $k_{min}$ and $k_{max}$ should be tuned to the approximation interval. Suitable positive real numbers $\epsilon_1$ and $\epsilon_2$ should be chosen.

If $Q_{min} < \epsilon_1$, then it is checked wether $Q_a < \epsilon_2$ applies (with $a > 1$, for example a = 2). If both conditions are met, then the iteration is finished.

If $Q_{min} < \epsilon_1$ is not met, $j$ is incremented by 1 (if a sufficient number of basis functions $\phi_{j,k}$ are chosen with respect to the approximation interval $I$).

If $Q_{min} < \epsilon_1$ is met but $Q_2 < \epsilon_2$ not, then $m$ should be increased.

**Remark:**

1) For the Shannon wavelet $j = 1$ was sufficient for most simulations. If steep slopes or large curvatures are present, good approximations where were calculated with $j = 2$ or $j = 3$. Here you can also start with a larger $m$.

2) Since $k_{max}$ and $k_{min}$ also depend on $j$ (i.e. for bigger $j$ a bigger $k_{max}$ and smaller $k_{min}$ is needed), with a bigger $j$ automatically a bigger $m$ should be chosen. You could double the value of $m$ when $j$ rises by 1. This rule could be useful in relation to the Shannon wavelet, taking into account the sampling frequency of the Shannon theorem.

3) Minimizing $Q$ instead of solving the equation system (2) has several advantages. One can use more collocation points and the least squares method is used to calculate the parameters $c_k$, because the differential equation is generally (if $y_j$ is not the exact solution) only approximately fulfilled (but the residuals are very small with good approximations). Moreover, the equations (2) several examples have been in ill-conditioned.

4) If $y$ has near the beginning big slopes or curvatures as with some stiff differential equations and only a good approximation in the interior of the interval $I$ is needed, then only $\tau_i \in [\tilde{t}_0, t_{end}]$ with $\tilde{t}_0 > t_0$ ($[\tilde{t}_0, t_{end}]$ is part of the overall approximation interval $I = [t_0, t_{end}]$) is sufficient for the calculation of $Q_a$. In this case the summation index in (5) does not start with $i = 1$ (e.g. $i = a$).

5) Although the Shannon wavelet does not have compact support, and no high order, but it returned in the simulations often significantly better results than other wavelet (even at relatively small $|k_{max} - k_{min}|$). In addition, it has several advantages for use in an approximation:

(a) The scaling function (as well as the wavelet) is defined analytically.

(b) The scaling function is many times continuously differentiable (see [11]).

(c) The scaling function is band limited and you can use this with the sampling theorem of Shannon, and thus "generalize" (see Remarks 1). This gives you information about the choice of $j$ in Fourier space.

# Comparing different wavelets

Finally we compare the approximation behaviour of different wavelets (Shannon, Daubechies of order 8, Meyer of order 3 and Battle-Lemarié of order 5). We minimize $Q$ and use the collocation points $t_i = i \cdot h$ (with $i = 1, 2, ..., m$; $m = r \cdot k_{max}$), with $h = 2/(r \cdot k_{max})$ and $k_{min} = -k_{max}$. It was $k_{max} = 15, 20, 25$, $r = 1, 2$ and $j = 0, 1, 2$ used.

Example 1: $y' = -t\,y$, $y(0) = 1$, $I = [-1, 1]$
Example 2: $y' = -2ty^2$, $y(0) = 1$, $I = [-2, 2]$
Example 3: $y' = -y - 2y^3 + \sin(2t)$, $y(0) = 0$, $I = [0, 4]$
Example 4: $y' = y - 2t/y$, $y(0) = 1$, $I = [0, 4]$

Example 5: $y'' = -y' - 257/4y$, $y(0) = 0$ and $y'(0) = 8$, $I = [0, 4]$
Example 6: $y'' = -100\,y$, $y(0) = 0$ and $y'(0) = 10$, $I = [0, 4]$
Example 7: $y'' = 3/2y^2$, $y(0) = 4$ and $y(1) = 1$, $I = [0, 1]$
Example 8: $y'' = 1/\zeta \cdot (y - (\zeta \cdot \pi^2 + 1)\cos(\pi \cdot t))$, $y(-1) = y(1) = 0$, $\zeta = 0.01$, $I = [-1, 1]$.
Example 9: $y'' = -ty'/\gamma$, $y(-1) = 0$ and $y(1) = 2$, $I = [-1, 1]$, $\gamma = 0.1$

We now compare the mean values $ln(Q_{min})$, $ln(Q_2)$ and $ln(mse)$. The mean values were formed over the logarithmic values.

| Wavelet | mean values | | |
| --- | --- | --- | --- |
| | $ln(Q_{min})$ | $ln(Q_2)$ | $ln(mse)$ |
| Shannon | -23.0799222 | -12.9773489 | -20.0139956 |
| Daubechies | -11.1367372 | 3.67751667 | -7.689269 |
| Meyer | -26.1509444 | -14.1988722 | -21.4952933 |
| Battle-Lemarié | -10.9775212 | 4.46392222 | -4.58829762 |

Table 4

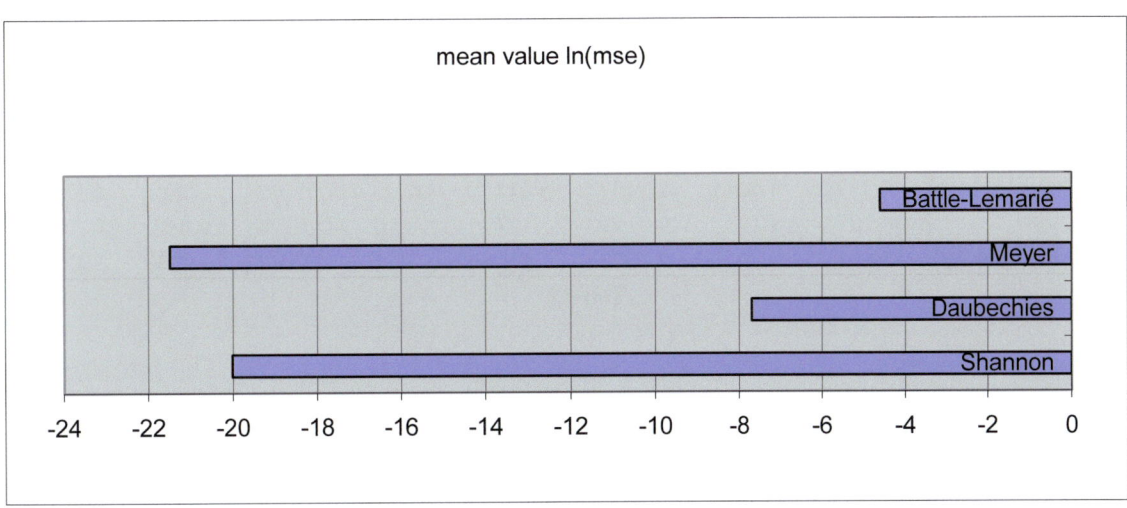

Figure 29

| ex. | Wavelet | Mean value | | | median | | | std. deviation | | |
|---|---|---|---|---|---|---|---|---|---|---|
| | | $\ln(Q_{min})$ | $\ln(Q_2)$ | $\ln(mse)$ | $\ln(Q_{min})$ | $\ln(Q_2)$ | $\ln(mse)$ | $\ln(Q_{min})$ | $\ln(Q_2)$ | $\ln(mse)$ |
| 1 | Shannon | -45.1416 | -34.9871 | -43.1020 | -59.5881 | -32.7114 | -40.7852 | 20.5402 | 18.4473 | 18.4535 |
| | Daubechies | -22.6407 | -8.5528 | -19.5461 | -18.2668 | -10.3142 | -19.8111 | 15.8459 | 5.6637 | 3.5565 |
| | Meyer | -44.5049 | -36.9365 | -44.6782 | -45.2432 | -33.3143 | -42.1668 | 24.3334 | 20.8302 | 20.6762 |
| | Battle-Lemarié | -24.7375 | -12.4429 | -19.7694 | -17.9303 | -12.3960 | -20.4020 | 16.1280 | 1.8081 | 1.9071 |
| 2 | Shannon | -23.3086 | -9.4866 | -17.4577 | -21.3585 | -10.0678 | -18.2280 | 12.5373 | 6.2265 | 6.7762 |
| | Daubechies | -7.6038 | 4.9098 | -5.9686 | -5.5510 | 3.2476 | -6.3710 | 6.9106 | 9.8605 | 4.7916 |
| | Meyer | -28.3080 | -10.2096 | -18.3243 | -25.9299 | -8.7265 | -16.5092 | 15.2511 | 7.5339 | 8.2264 |
| | Battle-Lemarié | -21.9864 | -4.1248 | -11.4590 | -17.3911 | -3.5546 | -12.0687 | 16.0386 | 4.0742 | 4.3339 |
| 3 | Shannon | -17.4368 | -7.4721 | -13.8434 | -16.9744 | -7.9037 | -13.6704 | 8.5688 | 7.2645 | 6.8050 |
| | Daubechies | -9.0168 | 10.3088 | -5.9482 | -7.4231 | 5.1840 | 1.2556 | 6.5367 | 16.0473 | 3.2873 |
| | Meyer | -21.6741 | -7.6675 | -14.3456 | -19.5319 | -7.8079 | -13.9251 | 14.4333 | 8.5846 | 7.4738 |
| | Battle-Lemarié | -24.3668 | -2.5878 | -9.5006 | -24.0618 | -2.4188 | -9.6513 | 19.0808 | 3.5698 | 4.0442 |
| 4 | Shannon | -24.1532 | -13.3252 | -8.4140 | -23.5640 | -12.3671 | -5.6234 | 16.6732 | 17.0482 | 12.6458 |
| | Daubechies | -17.6935 | 3.6399 | -5.4503 | -15.8065 | -1.1751 | -5.6441 | 7.8687 | 18.0739 | 4.2219 |
| | Meyer | -31.5287 | -16.8619 | -12.3544 | -34.2040 | -20.4032 | -13.4978 | 20.9125 | 13.2374 | 10.1363 |
| | Battle-Lemarié | -20.4299 | -4.5728 | -4.6614 | -13.4767 | -6.4869 | -4.8664 | 17.9377 | 7.5612 | 4.2059 |
| 5 | Shannon | -17.7301 | -3.4409 | -13.2128 | -15.2637 | 1.7691 | -9.2547 | 18.3200 | 13.0490 | 13.5674 |
| | Daubechies | -17.7301 | 3.4409 | -13.2128 | -15.2637 | 1.7691 | -9.2547 | 18.3200 | 13.0490 | 13.5674 |
| | Meyer | -22.7259 | -7.0328 | -16.9923 | -14.1805 | 1.0138 | -8.6092 | 20.2726 | 15.7955 | 16.1148 |
| | Battle-Lemarié | 0.2292 | 15.4572 | 1.0347 | 4.1586 | 15.1660 | -1.8023 | 14.3101 | 10.7499 | 5.7476 |
| 6 | Shannon | -10.7263 | 2.5601 | -6.1637 | -4.9058 | 7.1443 | -0.7929 | 15.9524 | 11.0973 | 11.4525 |
| | Daubechies | -18.0931 | 9.1515 | -0.6650 | -4.9914 | 9.0978 | -0.6859 | 26.1343 | 4.2960 | 0.0888 |
| | Meyer | -12.6006 | 3.2345 | -5.8314 | -8.4605 | 7.2359 | -0.8047 | 15.8750 | 8.4701 | 8.7088 |
| | Battle-Lemarié | -1.5164 | 16.3486 | 2.1872 | 4.6050 | 17.4862 | -0.4954 | 17.6942 | 10.6804 | 5.3346 |
| 7 | Shannon | -27.9718 | -23.2911 | -26.5562 | -30.3089 | -26.1877 | -33.1029 | 16.7338 | 14.7308 | 9.3988 |
| | Daubechies | -5.7594 | -1.0330 | -13.5146 | -6.6196 | -1.3947 | -15.2089 | 2.8753 | 3.3705 | 5.5848 |
| | Meyer | -29.1160 | -22.8944 | -26.5143 | -32.4942 | -25.6679 | -33.0689 | 17.6806 | 14.6209 | 9.6894 |
| | Battle-Lemarié | 3.2665 | 8.1615 | 1.1993 | 2.7752 | 5.0363 | 1.4432 | 1.9827 | 6.0327 | 0.7568 |
| 8 | Shannon | -17.7370 | -11.5636 | -24.4437 | -15.6891 | -12.7276 | -25.8691 | 14.2071 | 10.6355 | 15.2556 |
| | Daubechies | 0.6492 | 5.9883 | -3.0481 | 0.6996 | 4.8684 | -2.9766 | 0.2792 | 5.2747 | 0.2402 |
| | Meyer | -18.6732 | -11.1286 | -24.6974 | -19.0103 | -12.2142 | -26.4521 | 14.1214 | 10.0168 | 14.7596 |
| | Battle-Lemarié | -3.6220 | 14.8301 | -0.2754 | 5.7904 | 14.9184 | -0.5609 | 22.7281 | 3.8925 | 2.0306 |
| 9 | Shannon | -23.5139 | -15.7896 | -26.9325 | -17.0989 | -11.5982 | -20.8504 | 17.0451 | 13.2019 | 16.1509 |
| | Daubechies | -2.3425 | 5.2443 | -1.8497 | 0.0373 | 4.6431 | -1.7107 | 11.7608 | 4.5717 | 0.8279 |
| | Meyer | -26.2271 | -18.2931 | -29.7197 | -21.2537 | -11.4962 | -22.8726 | 19.2883 | 15.7809 | 18.4942 |
| | Battle-Lemarié | -5.6343 | 9.1062 | -0.0500 | 1.1396 | 6.9116 | 0.2970 | 18.4343 | 6.2239 | 0.8726 |

Table 5

What has been described here can be seen in Table 4 and 5 an Figure 29. The Shannon and the Meyer wavelet gave by far the best results for the nine differential equations. This was also reflected in other simulations with systems and examples from the reaction kinetics. In some examples the median over the logarithmic mean square error ($ln\ (mse)$) is even positive for the wavelets of Daubechies and Battle Lemarié.

# References

[1] Abdella, K. (2012). "Numerical Solution of Two-Point Boundary Value Problems Using Sinc Interpolation", *Proceedings of the American Conference on Applied Mathematics (American-Math '12): Applied Mathematics in Electrical and Computer Engineering*

[2] Ascher, U. A. Mattheij, R. M. M. Russell, R. D. (1988). „Numerical Solution of Boundary Value Problems for ODEs", *Prentice Hall (Series in Computational Mathematics)*

[3] Ascher, U. Christiansen, J. Russell, R. (1981). "Collocation Software for Boundary Value ODEs", *ACM Trans. Math. Software*

[4] Bertoluzza S. (2006). "Adaptive Wavelet Collocation Method for the Solution of Burgers Equation," *Transport Theory and Statistical Physics*

[5] Carlson, T. S. Dockery, J. Lund, J. (1997). "A Sinc-Collocation Method for Initial Value Problems", *Mathematics and Computation, Vol. 66, No. 217*

[6] Donoho, D. L.; (1992). "Interpolating Wavelet Transforms," *Tech. Rept. 408. Department of Statistics, Stanford University, Stanford*

[7] Hairer, E. Wanner, G. (1993). Vol. 1 : "Nonstiff Problems", *Springer 2. Auflage*

[8] Hairer, E. Wanner, G. (1996). Vol. 2 : "Stiff and Differential-Algebraic Problems", *Springer 2. Auflage*

[9] Mei, S.-L. Lv, H.-L. Ma, Q. (2008). „Construction of Interval Wavelet Based on Restricted Variational Principle and Its Application for Solving Differential Equations", *Hindawi Publishing Corporation Mathematical Problems in Engineering*

[10] Nurmuhammada, A. Muhammada, M., Moria, M. Sugiharab, M. (2005). "Double Exponential Transformation in the Sinc-Collocation Method for a Boundary Value Problem with Fourth-Order Ordinary Differential Equation," *Journal of Computational and Applied Mathematics*

[11] Qian, L. (2002). "On the Regularized Whittaker-Koltel'nikov-Shannon Sampling Theorem", *Proceedings of the Amarican Mathematical Society, Vol. 131, No. 4*

[12] Robertson, H. H. (1975). "Some Properties of Algorithms for Stiff Differential Equations", *J. Inst. Math. Applics.*

[13] Russell, R. D. Christiansen, J. (1979). "A Collocation Solver for Mixed Order Systems of Boundary Value Problems", *Mathematics of Computation*

[14] Schuchmann, M. (2012). "Approximation and Collocation with Wavelets. Approximations and Numerical Solving of ODEs, PDEs and IEs," *Osnabrück: DAV*

[15] Schuchmann, M. (2008). "Parameteridentifikation dynamischer Systeme auf günstigen Pfaden", *DAV*

[16] Schuchmann, M.; Rasguljajew, M. (2013). Error Estimation of an Approximation in a Wavelet Collocation Method. *Journal of Applied Computer Science & Mathematics, No. 14 (7) / 2013, Suceava*

[17] Schuchmann, M.; Rasguljajew, M. (2013). Parameter Identification with a Wavelet Collocation Method in a Partial Differential Equation. *Journal of Approximation Theory and Applied Mathematics (JATAM) Vol. 1*

[18] Schuchmann, M.; Rasguljajew, M. (2013). An Approach for a Parameter Estimation with a Wavelet Collocation Method. *Journal of Approximation Theory and Applied Mathematics (JATAM) Vol. 1*

[19] Shi, Z.; Kouri, D.J.; Wei, G.W.; Hoffman, D. K.; (1999). „Generalized Symmetric Interpolating Wavelets", *Computer Physics Communications*

[20] Strang, G.; (1989). "Wavelets and Dilation Equations: A Brief Introduction", *SIAM Review Vol. 31, No. 4*

[21] Unser, M. (1996). "Vanishing Moments and the Approximation Power of Wavelet Expansions", *Proceedings of the 1996 IEEE International Conference on Image Processing*
[22] Unser, M. Blu, T. (1998). "Comparison of Wavelets from the Point of View of their Approximation Error", *Proc. Of SPIE Vol. 3458, Wavelet Applications in Signal and Image Processing*
[23] Vasilyev, O. V.; Bowman, C.; (2000). "Second-Generation Wavelet Collocation Method for the Solution of Partial Differential Equations", *Academic Press*